Even in Chaos

INTERNATIONAL HUMANITARIAN AFFAIRS

Kevin M. Cahill, M.D., series editor

Even in Chaos
Education in Times of Emergency

EDITED BY **KEVIN M. CAHILL, M.D.**

A JOINT PUBLICATION OF **FORDHAM UNIVERSITY PRESS** AND
THE CENTER FOR INTERNATIONAL HUMANITARIAN COOPERATION
NEW YORK 2010

Fordham University Press has no responsibility for the
persistence or accuracy of URLs for external or third-
party Internet websites referred to in this publication
and does not guarantee that any content on such
websites is, or will remain, accurate or appropriate.

Library of Congress Cataloging-in-Publication Data

Even in chaos : education in times of emergency /
edited by Kevin M. Cahill.—1st ed.
    p.   cm.— (International humanitarian affairs)
"A Joint Publication of Fordham University Press and
The Center for International Humanitarian
Cooperation."
Includes bibliographical references.
ISBN 978-0-8232-3196–6 (cloth : alk. paper)
ISBN 978-0-8232-3197-3 (pbk. : alk. paper)
ISBN 978-0-8232-3198-0 (ebook)
1. Cognition in children.   2. Crisis management.
3. Chaotic behavior in systems.   4. Interdisciplinary
approach to knowledge.   I. Cahill, Kevin M.
LB1060.E96   2010
379.2′6—dc22

                                    2010005585

Printed in the United States of America

12 11 10   5 4 3 2 1
First edition

For my grandchildren

May your lives be full of dreams, and the joy of learning,
and of sharing your talents and productivity
with generosity, and, above all, with love.

CONTENTS

PART IV
Credo

The essays in this book highlight an inescapable fact: governments and the international community—the UN and non-governmental organizations included, have failed to ensure that the education of young people remains a priority—indeed, a fundamental human right—in the disruptive circumstances of man-made conflict and natural disasters. This book advances the international dialogue around this urgent need by identifying steps to protect our schools and ensure that they remain safe and nurturing environments even in the midst of the most difficult conditions. They point to the legislative strategies to combat the impunity of attacks on schools, students and teachers that has risen steadily in recent years.

The essays by experts in a range of fields share the urgent concern that Member States, the United Nations system and our NGO partners in humanitarian assistance must make education a priority in the response to complex emergencies. We need to develop a more coherent, rights-based response to these challenges and press for clearer resolutions, legislation, and policies to close this glaring gap in our policies and operations.

From my perspective as President of the General Assembly, I am most concerned about the recommendations that affect the policies of our Member States. As pointed out by the Special Rapporteur on the Right to Education, fully 90 percent of the countries where natural disasters and violent conflicts take place have governments that are unable to respond adequately to the humanitarian needs of their citizens, much less to maintain the safe havens of schools. They need much more assistance.

Government disaster-relief policies must integrate education into our humanitarian response and into the broader education framework as well. Education must be considered as important as food, shelter and medical attention. This requires cooperation and partnerships at all levels. I hope we will heed the call for more regional consultations and encourage Member States to explore this option with UN Regional Commissions, perhaps led by or with the participation of education ministers. I have no doubt that such meetings will facilitate the developments of better response and monitoring mechanisms at local, national, and global levels. They can also contribute in significant ways to improve data collection and analysis.

We must monitor peace agreements in affected countries to ensure that they too consider the integration and protection of schools and the educational infrastructure. Our planning must keep in mind that the restoration of the education system and poverty reduction must go hand in hand. And always, we need more reliable data on the conditions that are faced before, during and after crisis situations if we are to devise more effective response mechanisms relating to education.

As is so often the case, we have been reminded that the legal basis for the protection of the right to education is a part of much of our human rights legislation—from the Universal Declaration of Human Rights to the Convention on the Rights of the Child to the second Millennium Development Goal of universal primary education. But in the face of rising incidents of violence and human disasters cause by natural phenomena, these lofty goals fall victim to a culture of neglect, or even worse, impunity. Our collective failure to stop impunity serves as a license for the perpetrators.

I believe that government representatives are increasingly aware of the long-term value of protecting, nurturing, and educating their children in times of terrible trauma in their communities, and many of them are calling on the governments to take on greater responsibilities. We must urge all those involved, including UN bodies and agencies as well as civil society organizations, to develop clear policies that call on State Parties to protect schools and make them safe havens, especially in the most difficult situations. I support the call that States should criminalize attacks on schools as war crimes in accordance with the Rome Statue of the International Criminal Court and prosecute offenders accordingly. This must be done as a matter of course, routinely and systematically.

Our work is ever more urgent. While violent conflicts are more lethal, the increasing frequency of natural disasters affects seven times more people than violent conflicts. This is a trend that we must live with and we should apply all our tenacity and creativity to develop sound responses, ones that are feasible and that work in the worst of conditions.

Nearly forty years ago, when I was the director of Orbis Books, the Maryknoll publishing house, I worked closely with Dr. Kevin M. Cahill, publishing three of his books, *Medicine and Diplomacy*,[1] *Famine*,[2] and *Health and Development*.[3] I was impressed and inspired by his deep sense of responsibility to communities in dire needs of medical assistance. His commitment, first to people and then to institutions, pervaded the work of this intrepid epidemiologist who traveled to distant lands with unfading enthusiasm. His groundbreaking work provided our readers and me with powerful and valuable insights into the enormous challenges that face people in the developing communities and the importance that education plays in overcoming these challenges.

Today, decades later, I have continued to draw inspiration from Kevin's deep commitment to these communities and count him among my dearest friends. I have had the honor to have him serve as my Chief Advisor on Humanitarian Issues during my Presidency of the Sixty-third Session of the General Assembly. And it is in this capacity that he assisted my office in organizing a dynamic General Assembly dialogue on education in crisis situations.

I thank Dr. Cahill for bringing together the important contributions to this book and am confident that it will make us feel strong and more prepared in our commitment to better protect our learners, their teachers and all those involved in the delivery of their right to education. I believe these efforts have enabled us to make progress toward these important goals and see increasing awareness of right to education in emergencies as part of the legacy of my Presidency.

Father Miguel D'Escoto Brockmann
President of the Sixty-third Session
of the United Nations General Assembly
New York, 15 August 2009

ACKNOWLEDGMENTS

The editing of any multi-authored book is always a challenge. When contributors are preparing their chapters on many different continents, often in the midst of humanitarian crises, in a language that may not be their native tongue, and with a short deadline to assure that the volume can be published within six months in order to maintain the momentum of a global effort, the challenges are truly extraordinary. I express my deep gratitude to all of the authors for their generous response to my invitation to participate in this effort.

Dozens of individuals assisted in planning the U.N. General Assembly Debate on Education in Emergencies held in March 2009. The many organizations represented in the preparatory sessions for that Debate reflect the broad range of those concerned both about education in general and the rights of children in particular. Several particularly helpful to me were Abdel Rahman Azzam, Gina Bartolomeo, Christin Cahill, Peter Hansen, Ashley Hernreich, Ambassador Nassir Al Nasser, Chris O'Donnell, Alya Ahmed Saif Al Thani, and Caleb Zimmerman.

The Fordham University family supported this project from its inception; I enthusiastically express my gratitude to President Joseph M. McShane, S.J.; Senior Vice President/Chief Academic Officer Stephen Freedman; the Administrative Director of the Institute for International Humanitarian Affairs, Brendan Cahill; and the

Director of Fordham University Press, Fredric Nachbaur. Finally, Denis Cahill assisted in every stage of the long, arduous editing process and has my thanks and admiration for his literary skills and my love as his father.

ACBAR	Agency Coordinating Body for Afghan Relief
ADD	Attention Deficit Disorder
AFD	Agence Française de Développement
AI	Amnesty International
AIDS	Acquired Immune Deficiency Syndrome
AIMS	Afghanistan Information Management Service
ALNAP	Active Learning Network for Accountability and Performance in Humanitarian Action
ALP	Accelerated Learning Program
ALSS	Advanced Logistic Support Site
BEPC	Brevet d'Etudes du Premier Cycle (Junior High School Diploma)
BESE	Louisiana Department of Education Board of Elementary and Secondary Education
BSG	Brigade Support Group
CA	Civil Affairs Officer
CAAC	Children and Armed Conflict
CAP	Consolidated Appeals Process
CCA/UNDAF	Common Country Assessment/UN Development Assistance Framework
CDC	Centers for Disease Control
CERF	Central Emergency Revolving Fund
CESCR	Conventional of Economic, Social, and Cultural Rights

CFS	Child Friendly Space
CHAP	Common Humanitarian Action Plan
CIDA	Community and Individual Development Association
CIHC	Center for International Humanitarian Cooperation
CMOC	Civil-military operations centers
CNDD-FDD	Burundi rebel group: Conseil national pour la défense de la democratie / Forces pour la défense de la democratie
CNDP	Congolese armed group: Le Congrès National pour la Défense du Peuple
CPA	Central Provision Authority
CPN-UML	Communist Party of Nepal–United Marxist–Leninist
CRC	Committee on the Rights of the Child
CRED	Center for Research on the Epidemiology of Disasters
CRS	Catholic Relief Services
CWS	Church World Services
DAC	Development Assistance Committee
DDR	Disarmament, Demobilization, and Reintegration
DFID	Department for Foreign International Development
DOB	Deployed Operation Bases
DRC	Democratic Republic of the Congo
DRR	Disaster Risk Reduction
EC	European Commission
ECD	Early Childhood Development
ECE	Early Childhood Education
ECHO	European Commission's Humanitarian Office
ECOMOG	Economic Community of West African States Monitoring Group
ECOSOC	Economic and Social Council
EFA	Education for All
EMDH	Enfants du Monde–Droits de l'Homme
EPDF	Education Program Development Fund
ERC	Emergency Relief Coordinator
ESL	English as a Second Language
FAO	Food and Agriculture Organization
FECODE	Federación Colombiana de Educadores

FNL	Palipehutu–Forces nationales pour la liberation
FTI	Fast Track Initiative
GC	Geneva Convention
GEMAP	Governance and Economic Management Assistance Program
HDC	Humanitarian Dialogue Center
HEWS	Humanitarian Early-Warning System
HPN	Humanitarian Practice Network
HRR	Humanitarian Response Review
HRW	Human Rights Watch
IAPTC	International Association of Peacekeeping Training Centers
IASC	Inter-Agency Standing Committee
ICC	International Criminal Court
ICRC	International Committee of the Red Cross
IDF	Israel Defense Forces
IDHA	International Diploma in Humanitarian Assistance
IDP	Internally displaced person
IEEP	International Institute for Education Planning
IFRC	International Federation of Red Cross and Red Crescent Society
IHL	International Humanitarian Law
IIHA	Institute of International Humanitarian Affairs, Fordham University
IMF	International Monetary Fund
INEE	Inter-Agency Network for Education in Emergencies
IOM	International Organization for Migration
IRC	International Rescue Committee
IRIN	Integrated Regional Information Networks
JRS	Jesuit Refugee Service
LIC	Low Income Country
LRA	Uganda's Lord's Resistance Army
LTTE	Liberation Tigers of Tamil Eelam
MDG	Millennium Development Goals
MONUC	United Nations Mission in the Democratic Republic of Congo

MP	Military Police
MRM	Monitoring and Reporting Mechanism
MSF	Médecins Sans Frontières—Doctors Without Borders
MS: INEE	Minimum Standards for Education in Emergencies, Chronic Crises and Early Reconstruction
NACSA	National Association of Charter School Authorizers
NATO	North Atlantic Treaty Organization
NDMA	National Disaster Management Authority
NGO	Nongovernmental organization
NORAD	Norwegian Agency for Development Cooperation
NPA	National Plan of Action
NYPAW	Network of Young People Affected by Conflict
OAS	Organization of American States
OCHA	Office for the Coordination of Humanitarian Affairs
OCHR	Office of Civilian Human Resources
ODA	Overseas Development Assistance
OECD	Organization for Economic Cooperation and Development
OECD DAC	Organization for Economic Cooperation and Development–Development Assistance Committee
OEOA	Office of the Emergency Operations in Africa
OFDA	Office of Foreign Disaster Assistance (U.S.)
OHCHR	Office of the United Nations High Commissioner for Human Rights
ONUB	United Nations Operations in Burundi
OPT	Occupied Palestinian Territory
OSCE	Organization for Security and Cooperation in Europe
OSOCC	Onsite Coordination Center
OSRAC	Operations Safety Regulatory Action Committee
OTP	Open Training Platform
PA	Palestinian Authority
PACE-A	Partnership for Advancing Community and Education in Afghanistan
PCW	Protection of Women and Children
PKO	Peace Keeping Operation
PRS	Poverty Reduction Strategies

PRT	Provincial Reconstruction Team
PTA	Parent Teacher Association
RCD-G	Congolese armed group: Rassemblement congolais pour la démocratie–Goma
RSD	Recovery School District
SARS	Severe Acute Respiratory Syndrome
SCR	Security Council Resolution
SPLA	Sudan Peoples Liberation Army
SPLM/A	Sudan's People's Liberation Movement and Army
SRSG	Special Representative to the Secretary-General
SRSG-CAAC	Special Representative to the Secretary-General on Children in Armed Conflict
TANGO	Technical Assistance to NGOs
TLS	Temporary Learning Spaces
UDHR	Universal Declaration of Human Rights
UNAMID	African Union/United Nations Hybrid operation in Darfur
UNAMSIL	United Nations Mission in Sierra Leone
UNDAC	United Nations Disaster and Coordination
UNDP	United Nations Development Program
UNEP	United Nations Environment Program
UNESCO	United Nations Educational, Scientific, and Cultural Organization
UNGA	United Nations General Assembly
UNHCR	United Nations High Commissioner for Refugees
UNICEF	United Nations International Children's Emergency Fund
UNIENET	International Disaster Management Information Network
UNMIK	Interim Administration Mission in Kosovo
UNO	United Neighborhood Organization
UNPROFOR	United Nations Protection Force (in Yugoslavia)
UNRPR	UN Relief for Palestine Refugees
UNRWA	United Nations Relief Works Agency for Palestine Refugees
UPE	Universal Primary Education

USAID	U.S. Agency for International Development
USG	Under-Secretary-General
USIP	United States Institute of Peace
WFP	World Food Program
WHO	World Health Organization

Even in Chaos

KEVIN M. CAHILL, M.D.

The prioritizing of health care in or just after wide-scale emergencies is done by a process called triage. The goal is to identify and assist the viable and devote available resources and manpower to assure that as many as possible of those injured survive. In the medical sphere—and I know this best as a physician who has worked extensively in conflict, post-conflict, and disaster situations—trained personnel must make rapid decisions as to who can be helped, and in what order. Medical triage is based on clinical judgment, rules, and standards derived from the difficult experiences of previous disasters. The early phase of such triage operations are dramatic—the stuff of television shows and movies—and often profoundly unsatisfying. One prays for the calm, the time when a multipronged therapeutic approach can be possible.

That is when all the many contributors needed to address complex humanitarian crises begin to serve as full and equal partners. Post-conflict operations demand many skills and are successful only when there is full coordination. Every profession is tempted to deceive itself that in resolving crises their contribution is the most important. Medical personnel, maybe more than most, may erroneously view the healing arts as the *sine qua non* in collaborative efforts.

Unfortunately, education has often been considered an indulgence that can be postponed till the development phase of reconstruction. That approach not only fails to provide an essential and comprehensive right of education for all, but it also denies to innocent, vulnerable children in need the critical protection that schools, educational safe places, provide. Education in such settings can also impart life-saving knowledge required by parents and youngsters learning how to survive in new and dangerous environments.

To again use the medical model, once the immediate hemorrhage has been stanched, and basic supports implemented to sustain life, the good physician has the rehabilitation specialist and psychiatrist, among others, begin to address the inevitable complications of severe trauma. So too at the communal level, humanitarian workers caring for those caught in the maelstrom of life that is reality after conflicts and disasters, must simultaneously initiate multiple programs to promote a rapid return to normality, and to protect the young from further predictable assaults.

Education is now recognized as an essential ingredient in every conflict and disaster response. Here schools—and using the broadest concept of education—are critical. Education is, as contributors in this volume will attest, not only an expression of a basic human right, but represents the only proven path to growth, development and peace.

My own awareness of the significance of education in conflict, post-conflict and disaster situations was part of the humbling process of a Western-trained physician as he evolved into a humanitarian worker. I had been taught how to diagnose and treat individual patients. But specializing in tropical medicine brought early experiences—and there certainly were no lectures in my medical school or hospital training that prepared me for these challenges—in large epidemics and an appreciation of the plight of those caught in the cross fire of wars over which they had no control. Suddenly one had to learn how to deal with hundreds, thousands, and tens of thousands of women and children in dire straits, to be adaptable, and to function within very different cultures.

One quickly realized that individual diagnosis and therapy was a luxury one simply could not afford. One learned to coordinate food, water and sanitation programs, recognizing the public health imperatives of these nonclinical disciplines. In every refugee setting that I have known, the first sign of stabilization is the almost universal desire of mothers and children to establish a secure space for play and learning—the seeds of an educational operation.

Education is a manifestation of society's belief that somehow, someday, somewhere there will be a life after the near death that children experience in conflict and post-conflict situations. This book reflects the growing global commitment of the United Nations, its Member States, international and local non-governmental organizations, and concerned individuals that all children, everywhere, even under the most horrific conditions, have a right to education. The international community—with the leadership of many of the contributors to this book—has begun to establish better methods and minimal standards to assure that this too often neglected component of humanitarian assistance is recognized and supported.

Treaties and conventions and millennium goals regarding the universal right to education already exist, but funding and the political will to implement this noble rhetoric lag far behind. Trained teachers are often lacking, and in many circumstances, are an endangered species. Children remain the most vulnerable—and most exploited—in the chaos that inevitably accompanies conflict and natural disasters. But as Heraclitus observed thousands of years ago, opportunity does exist in adversity. Many of the chapters in this book demonstrate important lessons learned in war zones, and these greatly influence our current response.

Establishing the islands of tranquility that schools—even the most basic safe space devoted to the young—creates one of the most wondrous and satisfying sights in any refugee camp. To see children recapture innocence, learn together, and to hear them laugh again and play is one of the great rewards of hard, lonely humanitarian work. Schools provide the setting where feeding programs save lives; it is also where children are first made aware of the dangers of war zones—unexploded land mines and evil adults who may seek to abuse or conscript the young as child soldiers, sex slaves, or hostages for ransom.

Some of the authors in this book participated in a major Thematic Debate that was convened in March 2009 by the President of the

United Nations General Assembly's Sixty-third Session, Fr. Miguel D'Escoto Brockmann. I was fortunate to serve as the President's Chief Advisor on Humanitarian Affairs and helped plan, organize, and Chair the opening session of that Debate. This book reflects but part of the profound humanitarian legacy of Fr. D'Escoto's tenure in office at the United Nations. In planning a volume that will hopefully be used by students in academia, as well as by field workers in the midst of crises, I went beyond the UN Debate by inviting experts from around the world to write of their own experiences.

This book is another effort by the Institute of International Humanitarian Affairs of Fordham University to emphasize the indispensable role that education plays in crisis situations. This is the eighth volume in a series that addresses different aspects of disaster relief operations; the other books in the series are listed opposite the title page. The Institute's focus on the importance of education in humanitarian assistance is also evident in the over 1,400 graduates from 127 nations that we have trained. Fordham now offers a Masters of Arts in International Humanitarian Action (MIHA), as well as shorter courses in humanitarian negotiations, mental health in conflicts, disaster relief management, and other relevant topics that prepare the humanitarian field worker to become a teacher. The quality of teaching is as the very heart of what is needed if education, especially in emergency situations, is to succeed.

This book, like the human heart, has four linked parts: the voices of those in the forefront of the struggle to explain the importance of education in times of emergencies; the tools they must use to realize their goals; select places where the need for education in conflict and post-conflict situations can be appreciated: and, finally, a poetic reaffirmation of the noble mission of all the contributors—celebrating the innocence and resilience of children so they can grow into healthy, caring adults.

Every social movement begins with—and is sustained by—the passionate voices of those who see wrongs and want to right them. Agencies and organizations such as UNICEF, UNESCO, and Save the Children have been in the forefront of the global alliance that is promoting the right of children to education in emergencies. In this opening section, I invited some of the leaders in this international movement to offer their views; here are the "voices" of children and a psychiatrist, a Sheikha and a President, as well as the personal reflections of a humanitarian worker and a United Nations official.

My only regret is that the reader cannot hear, as I have, the sounds of these "voices," because only with the spoken word can one fully appreciate the depth of their commitment and wisdom.

1 | Ensuring the Right to Education

H.H. SHEIKHA MOZAH BINT NASSER AL MISSNED

In recent years, there has been a determined and almost diabolical process to undermine the right to education. The past decade was marked by a growing awareness of the importance of education, in particular following the World Education Forum's adoption in 2000 of the *Dakar Framework for Action—Education for All: Meeting Our Collective Commitments* and the emphasis on education in the Millennium Development Goals (MDGs). However, the same period has also witnessed continuing attacks on students, educators, and indeed entire educational infrastructures. Almost daily, we hear reports of deliberate attacks on universities and professors in Iraq, on schools and teachers in Afghanistan. Recently, the world witnessed the attacks on schools and universities in Gaza. In fact, the International Save the Children Alliance estimates that some 40 million of the 75 million children around the world who are not receiving primary education live in fragile, conflict-affected states.

Although the purpose of the MDGs is clear, it is obvious that the international community is not taking them as seriously as it should. We cannot help but be perturbed when we observe the discrepancies between agreed upon international laws and conventions and violent attacks on education in conflict zones. There can be no greater example of this incongruity than the onslaught in Gaza, with its deliberate and direct targeting of educational institutions.

Has education really been reduced to a political commodity? Is it merely a tool used by politicians to manufacture the consent of loyal supporters or to deprive societies of progress? We are concerned that through the targeting of centers of learning, entire societies are deliberately being denied of their right to progress. We must not let education be used as a political commodity, an instrument placed in the hands of power brokers. And we must undertake measures to defend the right to education and protect against any interruption in its delivery, including and most especially, during times of conflict.

We are concerned by the fact that children and families are being psychologically scarred and traumatized when the one place they thought of as a "zone of peace"—their school—becomes a target of war. We are concerned that if firm action is not taken to insulate education from conflict, depressed and desperate youth will find themselves plunged further and further into a cycle of violence. For if children are not in school, they are surely receiving their education elsewhere.

Therefore, it is with a sense of urgency that the world community must act and identify ways to ensure compliance with international law, safeguard against attacks on education facilities, and put an end to impunity of those who attack education, including teachers and students.

Every child has the right to education. This right is guaranteed by the Universal Declaration of Human Rights, the Convention on the Rights of the Child, and other man-made treaties. It is now time to talk straight and ensure that these rights are upheld.

2 | Protecting Human Rights in Emergency Situations

VERNOR MUÑOZ

Introduction

Some sixty years after the Universal Declaration of Human Rights[1] (UDHR), the commitment to realizing the human right to education has been a signal failure. It has seen the goals of Education for All[2] and the educational targets of the Millennium Development Goals[3] continually subsumed to the logic of economics, which in turn sees education as nothing more than an instrument of the market. We are all affected to a certain extent by this failure. For some, however, the consequence is a complete denial of that right.

For much of my mandate as UN Special Rapporteur on Education,[4] I have paid particular attention to groups of persons traditionally marginalized and particularly vulnerable to exclusion from education. In so doing, I have attempted to establish the causes and circumstances surrounding their exclusion and the challenges that must be faced in order to promote the realization of their right to education. It has become clear from this that there remains an urgent need to redouble efforts to safeguard the right to education for those people—especially children, adolescents, and youths—who are denied any possibility of attending school or attaining an education as the result, direct or indirect, of an emergency situation affecting their community.

For the purposes of this chapter, "emergency" refers to any crisis situation arising from natural causes (such as an earthquake, tsunami, flood, or hurricane), or to armed conflict, which may be international (including military occupation) or internal (as defined in international humanitarian law), or post-conflict situations that impair, interrupt, delay, deny, or impede the right to education. Such situations put people's health and lives at risk and threaten or destroy public and private assets, limiting the capacity and resources to guarantee human rights and uphold social responsibilities. Recurrent and/or combined emergencies in impoverished regions may of course have a multiplier effect, with devastating consequences for school infrastructure, teaching, and the educational opportunities generally of the children living in those regions.

Emergency situations are becoming increasingly frequent the world over.[5] However, the impact on each person directly involved in an emergency, while invariably brutal, may also vary, as will his or her personal reaction. At no time should such situations entail suspension of domestic and international obligations to guarantee the human rights of all those affected. State institutions, the international community, organizations, and individuals that offer assistance when they arise should be guided by those rights, rather than responding on the basis of often unwarranted and incorrect assumptions or financial risk. Further, for those that do offer assistance, they should act with those affected rather than for them.

Article 26 of the UDHR acknowledges that everyone has the right to education, which should be directed to the full development of the human personality and to the strengthening of respect for human rights and fundamental freedoms and further "promote understanding, tolerance and friendship." Yet we would do well to heed history and recognize that education can be of a kind that does not build peace but increases social and gender inequalities and may well fuel conflict.

Finally, there is a disjunction between social, cultural, and economic structures and educational activities carried out in times of emergency. There is an urgent need to close this gap because, although the impact of every emergency is different, there is one prevailing characteristic common to all: the interruption, degradation, or destruction of education and educational systems.[6]

Education in Emergencies

There is a multiplicity of proposed definitions and conceptions of "emergency" and the stages or time frames they reflect. The focus here will be on the period from early response to an emergency to the initial stages of reconstruction, for this is when what are perhaps the worst violations of the right to education occur.[7] It is during this period that educational systems and opportunities are destroyed and that the limited attention paid by the humanitarian agencies involved, and the relative absence of clear programmatic principles, indicators, or funding, are most clearly revealed.

The role and content of education in emergency situations are also a source of conceptual disagreement, especially where a distinction is being made between education in emergencies and education in non-emergency situations.

Context

The consequences of brutal armed conflicts and of natural disasters for education have become increasingly visible. Either can strike in any region, often without warning. No State is exempt; all have differing forms and levels of resources upon which they can draw to deal—or otherwise—with the consequences. In all of them, the civilian population is the chief casualty.

Statistics on conflict-related emergencies remain disturbingly vague, as most are based on estimates, which vary dramatically. In 2003 UNICEF stated that 121 million children were affected by armed conflict,[8] yet in 2000 UNESCO had put the figure at 104 million.[9] A comprehensive review in 2004[10] estimated the number of children and adolescents affected by armed conflict and without access to formal education to be at least 27 million, most being internally displaced persons (90 percent). More generally, approximately half of children who receive no education live in States where there is or recently has been armed conflict and where, in some States, net school enrollment is below 50 percent.[11]

The number of refugee and displaced children receiving no education outside UNHCR camps remains unknown, as does the number of illiterate young people, adolescents, and adults who have no educational opportunities.

Even though natural disasters are "statistically less lethal" than conflicts, causing one-third the number of deaths, in the 1990s natural disasters affected seven times the number of people affected by conflict.[12] Notably, natural disasters are on the rise, having occurred three times as often in the 1990s as they did in the 1950s. There are no reliable data permitting a comparison of the impact of natural disasters and the impact of armed conflict. There are reliable data, however, showing that around 90 percent of those affected by natural disasters live in States with limited capacity to cope with that impact.[13]

Statistics in themselves are not always sufficient in showing the degradation and destruction of education systems when an emergency arises, particularly in the case of armed conflict where teachers, students, and parents become the targets of violence. Parents keep their children at home to avoid the risks involved in the trip to and from school and to avoid falling victim to landmines.

The killing of students and teachers and the bombing and destruction of schools have escalated sharply over the past four years in terms of victims and brutality,[14] and in certain states, Afghanistan being a notable example, there is a clear gender dimension. Such attacks are directed against girls' schools, the sole intent being to intimidate and prevent girls from accessing education.[15]

Emergencies severely affect people with disabilities in particular. In her now well-known report *Impact of Armed Conflict on Children*,[16] Graça Machel noted that, for every child killed, three children are seriously injured or permanently disabled. More specifically, she found that armed conflict and political violence are the leading causes of injury and physical disability, and they are primarily responsible for the desperate conditions of more than four million children who currently live with disabilities and for the lack of basic services and/or minimum support.

The Importance of Education in Emergencies

Although I am personally opposed to the current tendency to treat education as no more than a tool, I recognize that, beyond the human rights imperative, education also provides physical, psychosocial, and cognitive protection that can be both lifesaving and life-sustaining. Education offers safe spaces for learning, as well as the ability to identify and provide support for affected individuals, particularly children and adolescents.

Education can also directly save lives by protecting against exploitation and harm, including abduction, recruitment of children into armed groups, and sexual and gender-based violence. In addition, it provides the knowledge and skills to survive in a crisis through, for example, the dissemination of lifesaving information about landmine and cluster bomb safety, HIV/AIDS prevention, conflict-resolution mechanisms, and peace building.[17]

Humanitarian aid traditionally focuses on the three classic areas of food, health, and shelter. Assistance, however, should be geared to people's overall needs and welfare, which, as noted before, clearly implicates education.

International Legal and Political Framework

The international legal and political framework of education in emergencies is the product of several global developments: the ever-increasing number of natural disasters, the changing nature of conflict and the fight against terrorism, and an unwavering perception of what education should be and the quality and kinds of education that should be available.

As parties to human-rights treaties, States have an obligation to respect, protect, and fulfill the right to education, whether or not an emergency situation prevails. In addition, the right to education inheres in each person regardless of legal status, whether "refugee," "child soldier," or "internally displaced."

Legal Framework

The UDHR establishes, in article 26, the right to free compulsory elementary education. Article 13 of the International Covenant on Economic, Social and Cultural Rights[18] defines the scope of this right more precisely, requiring that education should be available to all who have not received or completed primary education.

The Committee on the Rights of the Child (CRC) obliges States to ensure, without discrimination of any kind, access to education for all children living in their territories.[19] Its Article 28 promotes free compulsory primary education, urges States to develop accessible secondary education and other forms of education, and encourages international cooperation in educational matters.

Special attention must also be paid to the real aims of education, which are interpreted by the CRC as transcending mere access to formal schooling and embracing a broad range of life experiences and learning processes that enable children, individually and collectively, to develop their personalities, talents, and abilities and live a full and satisfying life within society.[20]

Moreover, under Article 22, States are obliged to ensure that a child who is seeking refugee status receives appropriate protection and humanitarian assistance and enjoys all rights as set forth in the CRC. Of particular importance is Article 38, which calls on States to respect and ensure respect for international humanitarian law.

The Optional Protocol to the Convention on the Rights of the Child on the involvement of children in armed conflict[21] has the potential to reduce the number of children recruited into regular armies and irregular armed groups and mitigate the implications for their educational opportunities.[22]

The accountability mechanisms of the CRC[23] remain weak, for they provide for no more than State party reports. Nonetheless, the CRC has shown a special interest in and commitment to, the issue of education in emergencies, as reflected in its guidelines for submission of reports, its recommendations, and its 2008 Day of General Discussion on education in emergencies.[24]

The Convention Relating to the Status of Refugees[25] provides that refugee children should be accorded the same treatment as is accorded to nationals with respect to elementary education (Art. 22, para. 1) and treatment no less favorable than that accorded to foreigners with respect to education other than elementary education. UNHCR found it necessary, however, to gear much of its work toward the protection of displaced persons, despite the lack of specific mandate within its statute for such work.[26] The growing number of displaced persons and the lack of specific legal protection prompted the development of the Guiding Principles on Internal Displacement,[27] on the basis of international humanitarian law and international human rights law.

International humanitarian law establishes a regulatory framework protecting the right to education during armed conflicts. The Geneva Convention Relative to the Protection of Civilian Persons in Time of War[28] states that measures should be taken to ensure that children who are orphaned or separated from their families as a result of a war have access to education.[29]

The 1977 Additional Protocol II[30] to the Geneva Conventions, applying as it does to non-international conflicts, is of the utmost relevance today as it covers the actions of non-State armed groups.

Of particular importance is Article 8 of the Rome Statute of the International Criminal Court,[31] which states that all intentional attacks on buildings dedicated to education constitute war crimes and are therefore subject to the Court's jurisdiction.[32]

International Political Responsibilities

The recognition given in Articles 4 and 28 of the CRC to the need for international cooperation in order to implement the right to education has not been translated fully and clearly into political responsibilities for the international community.

Nonetheless, the goal of education for all set up by the World Conference on Education for All,[33] held in Jomtien, Thailand, in 1990, certainly moved the language of human-rights obligations toward a future responsibility concerning the establishment of minimum standards in basic education. The Dakar Framework for Action on Education For All[34] was adopted at the World Education Forum,[35] held in 2000. The World Forum paid greater, albeit still insufficient, attention to the educational consequences of emergencies, placing special emphasis on children affected by conflict, natural disasters, and instability.

In contrast to political moves preceding them, the Millennium Development Goals[36] do not use the language of rights and State obligations. Instead, they assign educational goals to a development rather than a rights agenda. The effect has been to narrow the view of education to that of a quantifiable access to a full primary education that is free, compulsory and of good quality by the year 2015 (Goal 2) and the promotion of gender parity by the year 2005 (Goal 3).

In emergency situations, the obligation remains on States to ensure the right to education, even though they might lack the requisite will and/or capacity to do so. In recognition of this, a variety of actors— international NGOs, national and international agencies, and some donors—have attempted to shoulder this responsibility in part. Of these perhaps and more specifically, the Minimum Standards for Education in Emergencies, Chronic Crises and Early Reconstruction (INEE Minimum Standards)[37] developed by the Inter-Agency Network for Education in Emergencies (INEE)[38] stand out. These were drafted as a direct response to the neglect of education within humanitarian aid efforts.

The INEE Minimum Standards offer a harmonized framework of principles and paths of action to all actors who may be involved in the provision of education during emergencies, for them to coordinate their educational activities and, even more importantly, to promote the acceptance of responsibilities.

Despite the growing awareness of the need for delivery of education in emergencies and the progress made in doing so, there still remains an enormous gap between the legal and political responsibilities of the international community and its action and funding priorities.

Donors' Action and Priorities

Priorities for Action

UNICEF and UNESCO, the UN agencies that have assumed leadership for education in emergencies, are formally committed to the right to education. However, this commitment is not always matched by the educational strategies of large sectors of the international community, including other UN agencies, intergovernmental organizations, development banks, and private-sector and civil-society agencies.[39]

Although some progress has been achieved, especially with the creation of the Inter-Agency Standing Committee's Education Cluster, education as a priority in humanitarian assistance will continue to remain beyond reach until this priority is recognized by all, including, primarily, governments.

Donors

Humanitarian assistance is underfunded, barely receiving two-thirds of the sums identified as being needed and formally requested.[40] Consequently, when priorities are set, education in emergencies is not high on the list. In 2004, for example, only approximately 1.5 percent of the total humanitarian commitments were earmarked for educational programs.[41]

There is a steady increase in the literature covering the challenges relating to the financing of education in emergencies, a selection of which are highlighted herein. These clearly implicate the need for monitoring, evaluation, dialogue, and dissemination of best practices and innovations.

The challenges relating to education in emergencies most frequently discussed include the following:

- There is a lack of sufficient and suitable funding for education in general and the failure to honor formal commitments, despite the adoption of policies and the support of many donors who promote Education for All and the MDGs.
- There is a dominant paradigm of aid, based on the widely held premise that assistance is most effective in States with stronger policies and institutional adjustments.[42] Despite bilateral donors' emphasis on the importance of assisting the countries with the most pressing needs, such States—also referred to as emergency-affected fragile States—receive approximately 43 percent less funding than they would need based on the size of their population,

their degree of poverty, and their level of political and institutional development.

- Donors are reluctant to consider education as part of aid and humanitarian response, despite the fact that emergency situations can, and often do, last for many years.
- The priorities of donors have moved from the financing of long-term development needs to concentrating instead on humanitarian disaster relief. This frequently leads, as previously noted, to a focus on activities in the traditional fields of food, health and shelter.[43]
- There is a lack of continuity in funding between the onset of an emergency and reconstruction (often divided into "humanitarian phases" and "development phases").
- We have limited evidence concerning the effectiveness and responsibility of the providers of education in emergencies.

The limited involvement of donors in the implementation of the right to education has hampered coordination, the development of partnerships, examination of alternative funding models, and the building of risk-management capabilities.

Education Providers in Situations of Emergency

There is no single agency to which States requiring educational assistance can turn in an emergency. Neither is there a single funding mechanism for channeling financial resources. On the contrary, a plethora of actors take the stage, each with its own expertise, agenda and distinct priorities, mandates, capacities, spheres of influence, field presence, and financial bases. They include both agencies and other bodies of the UN system, bilateral and multilateral donors, international and domestic NGOs, and affected communities.

Education in emergencies enjoys a high level of awareness within the UN.

UNESCO has as its mandate to contribute to peace, security, and development through education and intellectual cooperation. A major effort that it has deployed since its foundation has been to ensure the right to education of persons affected by armed conflicts, through advocacy for a comprehensive understanding in the interests of peace. Although it has a wide-ranging mandate, UNESCO is painfully short of funds and other resources.

In general, the interventions of the agencies of the UN system are characterized by their concentration on primary education and by a concomitant lack of attention paid to tertiary education, particularly in fragile States.

If the agencies of the UN are to fulfill their mandates more completely, they will need to be adequately financed by the Member States. They will also need to revitalize their coordination efforts and raise the profile of the place occupied by education as a right in emergency situations.

Finally, although the World Bank has made important contributions to education in emergencies, it continues to work outside the human-rights framework. This reflects its strategy on education, which is to concentrate on support to education in the reconstruction stages following emergencies.

The Inter-agency Standing Committee's Education Cluster

The recent creation of the Inter-Agency Standing Committee's Education Cluster[44] is welcome, constituting as it does a first step toward the inclusion of education as a priority component of the humanitarian response.

The Education Cluster must act to meet the need to ensure a greater responsibility in that response on the part of the international community, including the UN, donor agencies and States, and local and international NGOs. It should become the proper mechanism for determining the educational needs in emergency situations and responding to them in a coordinated manner, for which purpose it should use and develop the tools laid down by the INEE.

Affected Populations

Refugees and Returnees

The educational options for this population are determined to a large degree by the repatriation efforts led by UNHCR. The aim of bringing about successful repatriation and reintegration of returnees, both teachers and students, has led to an emphasis in the study plans on all those aspects that recall the country of origin. This approach is not, however, always possible, as the relevant teaching material is often unobtainable or unsuitable. Such materials may be, for example, a version of the curriculum as it existed before the conflict, one that may even have contributed to the conflict itself, or it may be a mixture of the local model of the study plan plus innovations made to it by an NGO.

Internally Displaced People

Internally displaced people are disproportionately denied their right to education, with estimates standing at approximately 90 percent. This may be due to a number or a combination of reasons: ongoing lack of security, the lack of an international agency specifically mandated to respond to their needs, the lack of physical access for education providers, the lack of political will in governments to allow education providers to offer such people real opportunities, or the simple reluctance of governments to commit generally to fulfillment of the right.

Women and Girls

Gender parity in education is the focus of a global educational strategy that is obviously inadequate. In the context of emergencies, the relevant literature tends to concentrate on challenges other than parity: those created by the greater vulnerability of women and girls, including their problems of security, hygiene, and the lack of adequate sanitary facilities within the educational institutions, as well as the shortage of female teachers and the fact that girls are also required to do housework.

The impact of emergencies on girls is more serious, given that historically they have been the victims of exploitation and emotional and physical aggression, especially sexual aggression. For this reason, it is of fundamental importance for early response to emergencies to develop appropriate curricula that can be adapted to their particular needs and rights.

Child Soldiers and Combatants

It is estimated that around 250,000 boys and girls worldwide have been recruited to serve not only as soldiers, but also in the detection of mines, or as spies, messengers, and members of suicide missions.[45] A large proportion of international attention has been focused on their demobilization and reintegration, in line with international disarmament principles and the reintegration and demobilization standards laid down in the CRC.

A rights-based approach is needed to ensure that all educational programs deal with the multiple discrimination experienced by child soldiers, discrimination directed, among others, toward adolescents, minorities, and those with disabilities.

Formal and informal education, vocational training, and social-capacity building in general have been identified by many former child soldiers or combatants as essential to their long-term well-being,[46] and their prioritization should be a guiding principle for assistance offered.

People with Disabilities

People with disabilities, of either sex and of all ages, and in most parts of the world, suffer from a pervasive and disproportionate denial of their right to education.[47] In emergencies, however, particularly during conflicts and the post-conflict period, their right to receive special support and care is not always recognized by communities or States.

Young People and Adolescents

Governments and the international community have traditionally disregarded the education of young people and adolescents, since priority is always given to primary education.

However, there are an increasing number of experiments in accessible, realistic, relevant, and flexible learning, promoted primarily by international NGOs that offer youngsters an alternative basic education. These initiatives have largely been ignored by governments and donors, possibly owing to their lack of emphasis on standardization.

Consultations with Children

My various consultations with children and adolescents who have lived through conflict situations point to certain similarities in educational experiences and hopes. It is evident, for example, that conflict has a serious impact on their enjoyment of the right to an education that is free of charge, compulsory, relevant, and of good quality, especially for the children still living in the affected areas.

As many of the children and adolescents indicated, access to education and whether or not children remain in school depends to a large extent on the cost of education to them, including uniforms, teaching materials, food, and travel. They also voiced concerns regarding the extremely poor state of the school infrastructure, and some indicated that they had to walk long distances to reach school and in so doing were afraid of attack by armed groups.

Curriculum, Quality, and Shared Learning

The objectives of Education for All set out in the Dakar Framework for Action clearly state that access to a quality education is a basic human right of the victims of conflicts and natural disasters.

The quality of education implies a collective responsibility that includes respect for the individual nature of all persons; and it implies respect for and empowerment of diversity, since any learning demands the recognition of the other as a legitimate being.

The transition from emergency intervention to large-scale reconstruction provides unique opportunities for curriculum design and for improving the quality of learning. This requires generating data and minimum standards and proposes introducing innovative, flexible, and dynamic assessment systems.[48] The development of the curriculum and the wide spectrum of teaching activities that this includes require democratic and participatory attitudes in teachers and students alike.

In conflict and post-conflict situations, the new curriculum development that is required must be based on a detailed analysis and an understanding of any role played by the previous education system in, such that the emergency itself may turn into an opportunity for qualitative change.

An urgent task for governments has to be education for peaceful coexistence. Education for peace shares the same objective as human rights and should involve education as a whole, rather than as isolated components of the curriculum. It should make possible the understanding by all learners of the causes and consequences of emergencies.

Recommendations

The following measures should be taken so as to guarantee the immediate priority of this right:

- Greater emphasis must be placed on guaranteeing the right to education during emergency situations, in contrast to the current focus on postconflict situations.
- Increased action must be taken to bring to end the impunity for persons and armed groups, including regular armies, that attack schools, students, and teachers.
- There is need for further research into the effectiveness of some of the measures prompted by the increase in violence against schools, teachers, and students, such as armed responses in defense of communities and the promotion of resistance.
- Although there is an increased interest in the allocation and effectiveness of assistance in emergency situations, greater attention should be paid to assigning more resources specifically to fragile States.

- Prompt attention should be paid to the consequences of emergency situations for girls and female adolescents, and strategic measures developed to give physical and emotional protection in order to ensure their attendance at school.
- Increased and more thorough research is needed into specific programs for young people and adolescents, including the particular needs of persons with disabilities.
- Greater attention to understanding and the development of education for peace is required.
- There should be a shift away from the current emphasis on quantifiable, but often inaccurate, figures on, for example, school enrollment and dropout rates, and greater use of qualitative methodologies that will make it possible to determine the degree of psychosocial care required during emergencies.

Recommendations to States

- Develop a plan that prepares for education in emergencies, as part of their general educational programs, to include specific measures for continuity of education at all levels and during all the phases of the emergency. Such a plan should include training for the teachers in various aspects of emergency situations.
- Draw up a program of studies that is adaptable, nondiscriminatory, gender-sensitive, and of high quality, and that meets children's and young people's needs during emergency situations.
- Ensure the involvement of children, parents, and civil society in planning school activities, so that safe spaces are provided for students throughout the emergency.
- Design and implement specific plans to avoid exploitation of girls and young women in the wake of emergencies.

Recommendations to Donors

- Include education in all their humanitarian assistance plans and increase the education allocation to at least 4.2 percent of total humanitarian assistance, in line with need.[49]

- Actively support the Inter-Agency Standing Committee's Education Cluster.
- Use the Inter-Agency Network for Education in Emergencies Minimum Standards as a basis for the educational activities that are part of humanitarian response.

Recommendations to Intergovernmental and Non-Governmental Organizations

- Guarantee that educational responses to emergencies are in line with the INEE Minimum Standards.
- Seek mechanisms to ensure greater and more effective NGO involvement in the Inter-Agency Standing Committee, with a view to improving the coordination of the humanitarian response in the area of education.
- Organize and coordinate efforts for the effective implementation of quality programs of inclusive education during the emergency response.

3 | The Child Protection Viewpoint

ALEC WARGO

Although the subject of this book is substantially wider, I will limit myself to personal, field-based perspectives on the often fraught relationship between education[1] and child protection in armed conflict. I hope that this personal perspective, garnered from years working in the protection field, will remove us from the world of guidelines and policies and return us to the flesh-and-bone realities around the globe, where students, their teachers, and their communities often find themselves in the midst of armed conflict.

When Education Protects

At the start of my career in child protection, with UNHCR in Central Africa, I was only "theoretically" aware of the role of education in protecting children from harm and abuse during conflict and post-conflict situations, and that education personnel could serve as a linchpin in a community's ability to protect its children.[2] Perhaps I still held a somewhat jaded American view of education as sets of buildings with teachers who tried to fill our heads with bits of information in order to pass a series of exams that would place us in later life into a particular skill or field. Sometimes school challenged us, often it bored us, but it was always there, taken for granted.

It was only after I began to work in refugee camps in the early 1990s and later with UN Peacekeeping in the field that I began to truly realize that education—when children do have access to it—is at the heart of many efforts at documenting violations perpetrated against children during wartime. It is also, on many occasions, the primary interface for a human rights worker—the locus and the focus of child rights monitors and advocates' efforts to protect children from violence and abuse in times of utter chaos. This can be broken down into three categories of interest for the protection advocate: (1) education as an alert and as a bridge to response, (2) education and protection from recruitment, and (3) education and rehabilitation and reintegration.

Education as an Alert and a Bridge to Response

Rights advocates who monitor and advocate for protection of children have long recognized that alerts from affected communities are the most efficient way to identify violations against their children. In many cases it is the educationalist who alerts us to these grave violations. In many places, education authorities and teachers are among the most highly respected community members, and education itself has high value and prestige. Common practice has been that education personnel are often trained in the rights children have to freedom from harm, often a simplified version of the Convention on the Rights of the Child (CRC) and/or aspects of the Refugee Convention or other applicable legal instruments. The training is done with the expectation that it forms a base upon which action for proactive protection of children's rights can be grounded. This is a worthy exercise and has resulted in a number of interesting dialogues with education and community leaders in affected locales. Many of these dialogues have centered on the ability of the child to partake in decisions of his or her life choices, and, of course, a considerable amount of dialogue has been on the rights of girls and women in their societies.

Let us assume for the purposes of this chapter that child rights and their protection in wartime, especially against grave rights violations, are more or less agreed upon across time and cultures. The question then becomes how can we translate this knowledge, this "rights-based" approach, into action that responds to those violations? This action can be difficult, and sometimes almost impossible, in most conflict situations. When a violation has occurred, protection staff are asked by children's families or community leaders to undertake specific interventions. In the large majority of cases, these interventions fall into one of two general categories of action: (1) corrective response and/or (2) accountability response.

For example, in 2002, while serving as a child protection adviser to the UN peacekeeping mission to the Congo (MONUC), I was approached through the local Catholic bishop about a case brought to his attention by a group of teachers in a village some three days distant. The case concerned an impoverished single mother's only child, a fifteen-year-old boy who had been excelling against all odds at school. A rebel group active in the area had forcibly abducted him while on his way to school some weeks earlier. The immediate request was to remedy the situation: to assist in finding and advocating for the release of the boy as soon as possible. It was felt that an accountability response was not practicable at the time, because the rebel group in question was in complete control of the area and there were no NGO or UN staff based in the vicinity to prevent repercussions against the family.

In another example, from a similarly isolated community, a rebel soldier raped an eight-year-old girl in front of her home. The headmaster of the community's primary school brought the case to the attention of a local human-rights group, and subsequently the members of that group passed it on to me. In this case, the parents sought both a response to the girl's condition and an accountability response. The girl, along with her mother, was referred to a rape response center for treatment, HIV screening, and psychosocial recovery. The soldier was later tried for the crime, found guilty, and sentenced by a military court.

Teachers and education administrators can, and do, serve as protection alerts with human rights and child rights agencies. However, the crucial factor is whether or not that educator is not only aware of the violations and the child rights perspective but also alert to the need for both a response to and, in some cases, a remedy for those violations. In my experience, these protection alerts can only work if educators are linked meaningfully to advocates and actors who can assist them in accessing services and/or an accountability response.

Additionally, a large majority of violations in wartime happen far from major cities and towns, where these actors are normally based. Most might assume that national or provincial education authorities serve as a link from the village to the central or provincial level. But, alas, education authorities are the first to suffer rupture when state control is lost and are often sorely under-resourced in conflict-affected states.

The outlook would be bleak if we took such a restrictive view of "education." However, in most cases educationalists do succeed in making these links from far-flung regions. When we reexamine these links from the deep field to urban centers, it becomes clear that when we step back and take a broader view of the "education establishment," it is much larger than most would realize. When we take into account the non-traditional education actors in these countries, we see that communities of faith, NGOs, and civil society are all highly active actors in a network of educational activities, both traditional and non-traditional.

The educational establishment should be viewed not only as the remnants of a Western-influenced Ministry of Education apparatus, but also as the communities of faith, as well as local and international NGOs and UN partners that offer one of the most resilient and widespread monitoring and alert mechanisms available to child-protection actors today. These education networks are also one of the best-organized and widespread "bridges" to response. While in many places these networks are engaged as major partners for community-based protection, they are unfortunately not always utilized in a systematic fashion by the international community, including the UN system and donors. More must be done to support and more deeply engage this wider educational establishment as a key partner in protection response.

Education and Protection from Recruitment

Though much has be written about how education can serve as a preventative measure against everything from domestic abuse to gender-based violence, most of my personal experience with prevention of grave violations against children in times of conflict centers around the recruitment of children. It is estimated that approximately 250,000 children, boys and girls, serve as child soldiers around the world at any given time, and their identification and release continue to be a major field of activity for UN, NGO, and civil-society actors.[3]

Educational staffs are instrumental in alerting communities to the fact that underage recruitment is a violation of children's rights and can result in untold harm. They are also often an essential first line of defense in preventing recruitment. School-going children are much less likely to be recruited than those children who are not undertaking some kind of formal or informal education. Why are children who go to school much less likely to be recruited on the whole? Most analyses suggest that children who have access to school, in the large majority of conflict situations where there is no state-guaranteed access to primary or secondary education, are often from less vulnerable families. These are families who have the wherewithal to resist or avoid situations where their children might be directly exposed to armed actors. Their children generally spend the large part of their days in a structured environment, and the link between educationalist community and family can be quite strong. These families, when they do receive warning of impending danger, are often able to move away or use other coping mechanisms to reduce the risk of recruitment of their children.

More vulnerable families either have no means to move or are so destitute that movement would expose them to the direct risk of starvation, or other dangers equal to or greater than the threat of recruitment. Therefore, when we examine the risk of recruitment for school-going children, we appreciate that they are often less vulnerable socially and economically, are able to be alerted through educationalist/community/family lines of communication of impending troubles more readily and, when subject to danger, are often able to move to safety.

School often serves a community's barometer for trouble, and educationalists are often among the persons in the community consulted on how best to avoid danger. For example, in the South Kivu province of the Democratic Republic of the Congo (DRC), in 2002–3, the Congolese armed group Rassemblemett Congolaise pour le Democratie-Goma (RCD-G) began to move its recruitment and training centers farther from urban areas to the isolated island of Idjwi in the middle of Lake Kivu to avoid human-rights scrutiny of their child recruitment. The first message I received as the MONUC child protection officer of those recruitments was sent through a human-rights worker from a group of teachers on the island. Though the teachers had sent many of their students to the relative safety of the provincial capital of Bukavu to avoid this fate, many children from inaccessible areas were forcibly taken to the island, and a large number of families who could not afford to move from the area or send their children away were recruited. Most were not enrolled in school at the time.

In similar circumstances we ran into hundreds of children, some as young as ten years of age, associated with the Conseil national pour la defense de la democratie/Forces pour la defense de la democratie (CNDD-FDD) rebel group of Burundi, who had migrated from their training camps on the Burundi border with Tanzania to the Uvira/Fizi area of the DRC, where the group maintained a rear base. Most had spent at most only one to two years in primary school and recounted how the CNDD-FDD targeted them for recruitment because "they had nothing else to do." Most of the children came from broken or extremely poor households with no means to send them to school. Many were forced to work to support the family from the age of eight or nine years old. Almost none of them could read or write.

This socioeconomic argument is important to appreciate fully the direct relationship between access to education and the consequent access to protection from grave abuses. Indeed, the other end of this reality is that vulnerable children with no access to schooling sometimes seek out armed groups for various reasons. It is largely true that most child soldiers are unwilling or forced recruits. However, I have run into a not inconsiderable number of child recruits who willingly "volunteered" to fight with armed forces or groups.[4] In my experience speaking with hundreds of these child soldiers over the years, only a very few were enrolled in school at the time of their recruitment. As mentioned earlier, many stated that they had "nothing to do" and were often in dire economic straits or had suffered estrangement from their parents or caregivers. The deceptive offers of a monthly salary (rarely, if ever, paid) or the chance to feed themselves through pillage or extortion was attractive; and lacking any other alternative, many did "volunteer."

Unfortunately, it took these children little time to realize that their decision was not in their best interest—and months or years of suffering, exposure, and disappointment followed. Many lamented never going to school or not being able to finish school, and the promise of schooling under child demobilization programs led more than one child soldier to walk considerable distances to the nearest UN office or outpost to seek release. Making education, formal and informal, more widely available in communities at risk for underage recruitment should be more seriously considered. Education has a worth as a sign of opportunity and status among children in conflict-affected communities, and that value can, and must, be more deftly employed to prevent underage recruitment in the future.

Rehabilitation and Reintegration

In the area of children affected by armed conflict, the advancement of an educational response for children separated or demobilized from armed forces and groups is probably the most developed to date. Most children desire either formal or informal education or skills training upon release from armed groups, although access remains slow and patchy. The high value most former child solders place on education is an end or good in itself, but especially for the older children, it also forms part of their desire to "bring something back home" to their families and communities. My experience speaking with demobilized child soldiers about their next steps has always centered on aspects of what they can now do to become productive members of their family and society.

For the former child soldier returning to his or her community, well-structured education programs can mean a lifeline to "normalcy," a future and, in many cases, a hoped-for wage. The structure and community integrated aspect of these programs, if undertaken properly, can provide a good grounding for efforts to prevent re-recruitment or other abuses, either during the conflict or in the post-conflict phase. These challenges are now even more complex when we consider that the numbers of children demobilized *during* a conflict or immediate post-conflict phase are rising relative to those demobilized only after the conflict has come to a halt and stability has returned. This trend for release of children during conflict is welcome but it comes with its own problems and challenges. Chief among these is the re-recruitment of children who have been reintegrated, in some instances on multiple occasions, back into their communities. Education, the structure and alert functions for protection advocates and actors, remains a high value for successful reintegration and protection of children demobilized while conflict continues.

Educationalists also play a key role in managing the fears and expectations of communities who receive demobilized child soldiers. While many parents and communities naturally welcome their sons and daughters back with open arms, some armed groups use children against their own communities and this can cause deep-seated distrust and fear. The UN and its NGO partners have gone a long way to involve community members, including educational staff, in perceiving former child combatants primarily as victims and to convince fearful parents and neighbors that it is a child's right to benefit from reintegration assistance and that this can not be done without the active acceptance and participation of the community itself. Alas, each situation is unique, and the work of educationalists and their partners in war-torn communities is different depending on the circumstances prevailing prior or during release. This only points out that much-needed research and analysis are required to further strengthen this approach.

To some extent, victims of sexual violence follow similar trajectories during their reintegration, though often their interface with education is concealed or is purposefully blurred to prevent the labeling of these boys and girls as sexual victims. This is not to say that educationalists do not undertake or have the potential to lend their support to these children. Indeed, in a number of circumstances known to me, these staff members have been a crucial support to these children and their families.

When Education Fails to Protect

As I have observed, education is at the forefront in preventing, or responding to, abuses against children in wartime. However, in modern warfare there is a palpable trend in which the educational personnel themselves, and/or the buildings in which that education takes place, become the objects of abuse or play a role in the abuse of children.

Attacks Against Schools

Most rights monitors would identify the 1990s as the watershed epoch, when attacks against schools gained the notice of child protection staff. We saw this in the activities of the Lord's Resistance Army (LRA) in northern Uganda and in southern Sudan, during the Rwandan genocide, and certainly in Liberia and Sierra Leone. We currently witness these attacks in places as varied as Nepal, Afghanistan, Pakistan, the DRC, Somalia, Palestine, and southern Thailand.

It is easy to oversimplify the commonalities of these attacks. Some were related to wider crimes, such as the genocide in Rwanda, where children and educated men and women, including teachers, were brutally murdered by gangs, many times in schools and churches. The perpetrators were intent on the destruction of a minority community and its perceived sympathizers or, later on, in revenge killings against those who were presumed complicit in the killings due to their ethnic affiliation. It is here that we find education personnel complicit in killings based on ethnic identity. Similar stories were recounted to me when I served as human rights officer with the Organization for Security and Cooperation (OSCE) mission in postwar Bosnia. Sadly, the very pillars of a community, who can give voice to protection and reconciliation, can also stoke the embers of ethnic hatred. This is a constant danger in societies in conflict where ethnic identity is a key factor, and the educational system, in both its personnel and as a packaged set of ideals and perceptions of the world, often stand at the very heart of it.

There are other instances when education and identity politics have led to massive rights violations and where schools themselves become a central battleground. I think of the recent conflict in Nepal and the politicization of Nepali education, which will haunt that country for some time to come. During that insurgency, the Maoist youth wing concentrated its recruitment efforts on rural schools throughout the country, abducting whole schools for days of Maoist "cultural" programs. These programs funneled healthy recruits as young as fifteen years of age to fighting units and struck fear into the hearts of teachers and education administrators—who had no choice but to look on or risk harm to themselves and others.

In the post-conflict phase in Nepal, the Maoist youth continue to engage in violent acts, both in schools and in their communities. Currently other political parties have followed suit and created their own youth wings, the largest of which is the Communist Party of Nepal–United Marxist-Leninist (CPN-UML). These youth wings are nominally under the control of the parties, and continue to be a source of inter-factional violent acts and killings. Their presence and propagation in schools across the country remain of deep concern to those assisting the peace process in Nepal.

Likewise, in southern Thailand, identity politics in the traditionally Malay areas has taken on a violent character. The instruction of Thai language and culture has become a lighting rod for local resistance by armed actors. Students and teachers in these schools, many of them of Malay identity, have been killed in shootings, bombings, and other attacks by armed groups. More worrisome is the effort on the part of Thai authorities to protect these schools by sending military forces to guard them. Though it may be well-intentioned, one must question the wisdom of deploying military personnel into schools, effectively making them targets and removing the civilian nature that such centers of learning must display.

Increasingly, there are also the depressing ideological battles fought over the very fundamentals of education. The questions of what can be taught and who can attend school have generated some of the most gruesome attacks against schools and students in recent history. The worst example of this is the current situation prevailing in the insurgent-controlled areas of Afghanistan, and the neighboring border areas of Pakistan, where the Taliban and their allies have wreaked havoc. The crux of the battle concerns two types of ideological issues: (1) are girls fit to attend school at all, especially in the company of boys? and (2) should children be taught anything beyond Taliban-approved interpretations of religious texts? It has now become commonplace for the Taliban or associated leaders to leave "night letters" threatening attacks on schools if they do not cease allowing girls to attend school or alter the school curriculum to reflect their conservative interpretation for educating young people. Girls' schools in particular have been attacked with bombs, poison gas, and a barrage of threats, and there is little one can do in the more isolated areas until some semblance of security prevails. Add to this the depressing fact that a significant number of madrasas in the border areas appear to be training children to undertake armed conflict as fighters or, worse, as suicide bombers.

Infiltration of Education by Armed Groups

When we survey the challenges presented when armed groups attack schools and the difficulties these acts present for protection of education, they pale in comparison to the infiltration of education by armed groups. This does not refer to the abduction or harassment of students and teachers described earlier, but the much more pernicious infiltration of education by armed group cadres and their sympathizers. The latest example of this is the Palipehutu–Forces nationales pour la liberation (FNL) in Burundi, which, in the period prior to the final peace settlement, demonstrated its hold over sympathizers in the educational establishment, resulting in teachers actively recruiting children for the FNL. It is believed that the recruitment was aimed at securing these FNL sympathizers a demobilization package, which, in turn was promised to the children, although children have been purposely excluded from such schemes in the peace process. Many of these children were exposed to militarized camps for fairly long stretches of time, up to two years, before the FNL agreed to their disqualification. We can also not discount the fact that, if the FNL had not come to an agreement with the government of Burundi, these children would have had a very real chance of taking up arms.

Similarly, in the Congolese refugee camps in Rwanda in the latter half of the 1990s and until quite recently, teachers and community leaders in those camps, sympathetic to Rwandan-backed rebel groups operating in the Kivu provinces of the DRC, actively recruited children to fight the Congolese government forces. As the child-protection focal point for UNHCR at the time, I was amazed that our data showed that an entire portion of the camps' population (boys aged from thirteen to early adulthood) had effectively disappeared. We knew from the data and corroborating evidence that the boys had been sent for military training or other support roles to either the RCD-G, the Banyamulenge (an ethnic Tutsi group residing in the mountains of South Kivu Province of the DRC) forces, or, latterly, the Congres National Pour la Defense du Peuple (CNDP). In this instance, teachers and community leaders were either complicit or were too frightened to give evidence. This recruitment is a worry in a number of refugee camps, where similar pressures exist and more must be done to monitor and halt such practices.

The Missing Children

Children missing due to recruitment, "voluntary" or forced, are a much smaller number than the very real problem of the thousands of children, the most vulnerable, who will never have the opportunity to go to school. In most "failed" or "failing" states the number of out-of-school children can reach 50 percent and more in the most isolated and poor regions, where conflict usually flourishes. These are the children whom the education establishment certainly fails to protect.

We have shown that education can work as an effective monitor and alert for human rights workers and as a bridge to response. But that very same system falls absolutely flat for these "missing" children, because education never sees their faces. It was not surprising then, that the majority of the children whom I interviewed upon release from one armed group or another recount how they were not able to attend school, of impoverished parents powerless to flee or to bribe their children's recruiters to allow their children to remain and who are also unable or unaware of the alert function that many educationalists can play. In any case, many of these families live days away from the nearest school or do not feel they are able to share their problems with persons with whom they are unfamiliar.

If education is to protect all children, and the CRC, the most widely adopted piece of international legislation, states that it should, then the international community must do much more to find ways to expand the education safety net during times of conflict. This could be achieved through the establishment of a range of formal to informal educational outreach and community sensitization programs. It should most certainly be prioritized in failed and failing states that are in crisis or conflict.

When Education Itself Needs Protection

From the perspective of the Office of the Special Representative for Children and Armed Conflict (SRSG-CAC), one overwhelming question has to be addressed that is of a different nature from the technical discussions on how to ameliorate the condition or respond to violations against children: How do we stop these attacks? How do we break the cycle of violence against children and their teachers in classrooms during conflict?

We are currently observing what current Special Representative Radhika Coomaraswamy has termed the "changing nature of conflict." And, unfortunately, its trajectory is not favorable to children. In the past children were considered peripheral to the conflict. Now they are often at its very center. Though many experts on this matter may argue why and exactly when this phenomenon of targeting of children and education first appeared, it is undeniable that it is now a fact and that these attacks are not abating. In fact, they are becoming more commonplace.

Again what to do? One word: accountability. Ninety percent of the armed actors cited in the Secretary-General's annual report on children and armed conflict are non-state actors. Some of these non-state actors have proxies that represent their interests on the sidelines of UN events such as the Human Rights Council and the Committee on the Rights of the Child. But the weight of the UN's state-oriented system is generally not sufficient to engender compliance by these parties.

Security Council Engagement

It was with this in mind that the former SRSG for children and armed conflict, Olara Otunnu, began a process in 1999 of engagement with reporting to the UN Security Council, positioning children and armed conflict directly on its peace and security agenda. This was the first "thematic" protection mandate to be entertained by the Council in such a way. Things progressed slowly at first, with annual reports of the Secretrary General speaking of a vague set of violations against children in times of conflict but without a systematized information gathering network to backstop claims of abuse.

The next breakthrough came in 2001 with Security Council Resolution (SCR) 1379 asking the Secretary-General to prepare an annual list of parties to armed conflict who recruited and used children under the age of eight years in armed conflict, the "list of shame." After this point the Secretary-General also began to call, in a more focused way, for compliance and, when lacking progress, for the possibility of the Council using its power to take measures, including sanctions, against groups and individuals who recruit children.

However, actual compliance was slow. This was, first, because the UN system was slow to engage in a concerted campaign in a unified fashion. Second, the information available to the Security Council at the time did not meet the rigorous requirements usually associated with the application of sanctions. Third, compliance work lacked a recognized and agreed format. This was rectified over the next two resolutions agreed by the Council. SCR 1539 of 2004 established the concept of a concrete time-bound "action plan" to halt the recruitment and use of children. These action plans called upon the UN country teams to engage with both state and non-state actors to agree to a set of measurable and time-bound activities to prevent recruit-ment, release children associated with those groups and verify compli-ance. Failing this, the Security Council reiterated its intention to consider the use of measures, including sanctions, against violating parties. At the same time the Council asked the SRSG-CAC to develop a system-wide plan to strengthen monitoring and reporting on grave violations against children in times of conflict. This was aimed at es-tablishing a more solid basis of information in which the Council might deliberate before exercising its power.

Information Is Power

The Secretary-General's plan to provide "timely accurate, reliable and objective information" to the Security Council was unveiled in his report (S/2005/72) of 2005. The plan is important because it de-fines the way in which reliable information should be collected, veri-fied, and packaged for the Security Council and also identified the six grave violations—including recruitment and attacks on schools—enumerated earlier in this chapter. It also reiterated a call for action plans and an intention to utilize all tools at its disposal to engender compliance among parties engaged in grave violations against chil-dren in armed conflict. Most importantly, it also set in place a Work-ing Group of the Security Council that would deliberate throughout the year on specialized reports of children affected by armed conflict, in situations listed in the Secretary-General's annual report, and make recommendations to the parties concerned as well as to the Council sanctions committees.

This plan as set forth by the Secretary-General and endorsed by the Security Council resulted in the first systematic monitoring, reporting, and verification mechanism on child protection. It identified head-quarters responsibilities, but it also mandated UN country teams to put in place task forces to undertake systematic monitoring and re-porting at country level. Since that time the Office of the SRSG-CAC has worked to mainstream monitoring and reporting throughout the UN system and technically backstop the work of UN country task forces on the design and implementation of action plans. It should be noted, however, that though the Council endorsed the monitoring of six grave violations, including attacks on schools, the action plans remained limited to the halt of recruitment and use of children as soldiers. Both monitoring and reporting (MRM) and the implementation of action plans advanced slowly but have now accelerated to the point where knowledge and expertise on MRM and the design and implementation of action plans to halt recruitment and use are widespread.

At the same time, the Council recognized the need to respond, with a recommendation to parties, on the other five grave violations listed in SCR 1612. The crucial lesson learned was that more could be expected of the Security Council Working Group regarding concrete recommendations when more in-depth analysis of the violations were available to them. Information is power, and it can and should be used by protection actors to great effect with this mechanism.

However, with a few notable exceptions, the information available on attacks against schools is largely spotty or absent in the Secretary-General's reports on country situations to the Working Group. Circumstantial evidence suggests that protection partners have not spent as much time examining the issue and empowering their partners to report and make suggestions on how to better protect schools and students during armed conflict. Without the information and analysis needed, attacks against schools and students and the power of the Security Council to compel parties to respect and protect education in conflict will remain woefully underdeveloped. Other violations have similarly languished, but advances *have* been made that might prove useful for partners wishing to strengthen the protection of education in conflict through this mechanism.

In July 2009, in its resolution 1882, the Security Council expanded the listing criteria from recruitment and use of child soldiers to include parties that kill and maim in contravention of international law, as well as parties that commit grave sexual violence against children in wartime. This is important for the additional attention and focus it will generate for country teams dealing with these violations, and it will also result in mandatory action plans to cover these violations. Much work remains to be done, both at headquarters and in the field, in response to these new challenges for protection, and it must include reaching out in a more concerted way to gender and human rights partners in the field.

As stated earlier, it is expected that, when violations are identified, country teams can suggest Security Council interventions as well as actions the parties should take under applicable national and international law. There is always the promise that the Council may mandate the UN and its country teams to expand the listing criteria and demand action plans on attacks against education at a later date. Human Rights Watch called for such a mandate during the Council's deliberations on SCR 1882. But for this to happen, UN specialized agencies and their partners have to originate and sustain efforts at better monitoring and reporting on attacks against education. Additionally, joint MRM task forces should utilize monitoring and reporting on attacks on education to strengthen their advocacy and recommendations to the Security Council Working Group on CAAC. Information is power. Its efficient delivery to those who can hold accountable violators is an opportunity that we can no longer afford to ignore.

Conclusion

I wish to reiterate that more can be done to protect education from attack in all its manifestations, to improve the ability of education to protect children, and to strengthen MRM to protect education in armed conflict.

With this in mind, I would make the following recommendations to the UN system, protection partners, educationalists and donors:

- Education is a key stabilizer and must be a part of any emergency planning in conflict prone of conflict-affected areas. Education's role in protecting children from abuse and harm through either direct action or alerts to protection personnel cannot be overstated.
- Protection should be seen as a crucial part of education in conflict-affected countries, and funding and training for protection should be built into any programs for these states.

- SCR 1612 and 1882-mandated monitoring and reporting can serve to better protect education during conflict, and UNESCO, educational agencies, and NGOs involved in education can and should join country task forces to better monitor attacks on education and to advocate with the Council on appropriate actions.
- Education outreach for vulnerable communities pre- and post-conflict is an important protection tool. Donors and agencies planning for child protection must seek to ensure broader coverage of disadvantaged communities.

4 Donor Investment for Education in Emergencies

BRENDA HAIPLIK

Around the world, there are approximately 75 million children out of school. More than half of them—40 million—live in conflict-affected fragile states.[1] In addition, an estimated 750,000 more children and youth have their education disrupted or miss out entirely on education each year owing to humanitarian disasters. Children experiencing emergencies are among the most vulnerable on the planet. These young global citizens often suffer not just from a lack of access to quality education but also from limited access to safe drinking water, sanitation facilities, and good nutrition. Some children may be exposed to HIV/AIDS and to other protection-and health-related threats. These figures and facts should be a major source for international concern.[2] If the global community is to meet the Millennium Development Goals and achieve Education For All by 2015 increased attention and action must be invested in education in emergencies.

There are numerous compelling reasons why donors should invest in education in emergencies. In the immediate aftermath of a disaster, natural or man-made, education can:

- Help protect children from death or bodily harm (schools as safe, child-friendly spaces)[3]
- Impart critical lifesaving information on simple hygiene and health issues and/or the dangers of unexploded ordnance
- Decrease children's vulnerability of being recruited into armed groups or being trafficked

- Reduce the effects of trauma and offer children, their teachers, and other affected education personnel and entire communities a sense of normality, structure and, perhaps most importantly, hope for the future.

Education is what children and their families demand. Evidence from a variety of emergencies, both man-made and natural, in different cultural, religious, and political settings around the world confirms that parents, children, and communities desire access to quality education as early on in an emergency as possible.[4]

Over the longer term, quality education can:

- Be a critical ingredient in the reconstruction of post-conflict societies
- Promote conflict resolution and peace, social cohesion, tolerance, and respect for human rights
- Increase children's future earning potential, enabling them to keep their families healthier and improve their ability to break out of the poverty cycle.

Although international awareness is growing of the potential value of education as a source of protection during emergencies and as a key to sustainable development many donor governments still categorize education as a sectoral programmatic response most suitable to the development stage.[5] Education, unlike life-saving health and water sectoral responses in emergencies, is generally not seen as an ongoing process that should continue, without interruption, even when an emergency, natural or man-made, strikes.[6] Until recently, education has been considered as only life-enhancing and not as life-saving in nature, hence a secondary-tier sector in emergency response.

Consider the following:

- Less than 2 percent of humanitarian emergency aid was directed toward education in 2007.
- Only 27 percent of the Global Education Cluster's funding requirements were met in 2007 (by Denmark, Ireland, Norway, and Sweden).

- Short-term education projects were funded more often than not as part of a longer-term continuum from emergency to development. (In mid-2009, only a handful of donors had policies that support education in emergencies: the Netherlands, European Commission (EC), Canada, Denmark, Japan, Norway, and Sweden).[7]

To add to these existing funding challenges, in the current difficult and prolonged period of global financial crisis donor governments have become even more critical in re-evaluating how they can better support potential recipient nations.

Quality Education: A Return on Investment

How can UN member states, donors, multilateral agencies, and organizations ensure a good return on their investments in education in emergencies? The link between quality education and a return on investment must be made clear. In emergency contexts the connection between access to and the quality of education is significant. "Surely, one cannot have quality without access, but access without quality is also meaningless. Without quality children will drop out of school."[8] As in non-emergency contexts, the value of education during emergencies is determined by the relevance of the curriculum to children's lived experiences, the quality of teaching and learning materials, the availability of trained teachers and the potential for post-primary education and/or employment opportunities. In emergencies, as in "normal" contexts, parents weigh the benefits of sending their children to school against the opportunity cost. If parents are expected to send their children to temporary classrooms during a challenging emergency period, then quality education must exist. Donor support, both monetary and technical, must therefore extend beyond the traditional acute phase of the initial emergency response (access to education, supplying teaching and learning kits and temporary shelter) and continue on through to the recovery and reconstruction phase and into development (quality education). To get a high return on investment, donors must work closely with education personnel in each and every emergency response to strategically meet the needs of a particular learning community, both in terms of access to and quality of education. Because there is no blueprint for effective response, funding should be flexible and needs-based. If a situation changes, original plans should allow for timely adaptation. By coordinating closely with organizations working in education on the front lines, donors can help to determine how funds can be most effectively utilized during every stage of the response, hence increasing the return on investment at all stages.

In purely economic terms, investing in education before an emergency unfolds is more cost effective than waiting until an emergency strikes. Data, research, and analysis on disaster-risk reduction, school safety, and past emergency response evaluations show that every dollar invested pre-emergency during the preparedness phase is equivalent to four dollars invested once an emergency is underway.[9] Educationists and donors realize that comprehensive, longitudinal research is required to better understand and document the long-term benefits of investing in education in emergencies.[10] In terms of the inclusivity of education, the Inter-Agency Network on Education in Emergencies (INEE) asserts, "It is less expensive if we incorporate approaches to support everyone at the outset of an emergency response, than if we try to change exclusionary school infrastructure and practices at a later date."[11] Identifying the most vulnerable learners (for example, children with disabilities, girls, children suffering from trauma) before an emergency takes place and including them in programming from the outset is most desirable.

Quality education is also a proven ingredient for sustainable peace, as highlighted at the Sarajevo conference (February 2009) on the connection between education and peace. At country level there are many examples of the key role that education plays in the peace process. In the current and historic post-conflict era in Sri Lanka, education cluster members are working to build peace education and social cohesion at the individual classroom level. Observes an International Save the Children Alliance (ISCA) report, "If the education programming is well-designed and the education activities provided are of high quality, schools can provide an entry point for encouraging conflict resolution, tolerance, and respect for human rights."[12] Quality education in emergency settings resulting in meaningful learning can be transformative in nature and radiate far beyond school walls. In programmatic terms, there are excellent resources and tools which have been developed, and are continuously used and updated, by practitioners who are learning as they go about what works and what does not work in real-life, complex, post-conflict emergency settings. These resources range from needs assessments, teaching and learning kits for particular populations (for example, students of alternative/accelerated learning programs), to strategic planning, monitoring, and reporting tools. The quality of education in emergency response, particularly in post-conflict settings where "healing" classrooms become zones of re-building peace, is enhanced by the use of these tools and resources. Within the global INEE network and beyond, resources and experiences are shared widely both online and through face-to-face workshops and seminars. This continuous sharing and evaluation contributes to the further refinement and increased quality of not just the resources themselves but of the effectiveness (higher return on investment) of future responses.

Investing in "Success": The Minimum Standards for Education in Emergencies

"An emergency response offers space to look at education with a fresh perspective. It can be an opportunity to improve on the previous standard of education provision and to address issues that had not been considered before," notes the INEE.[13] There are several key elements through which the global education community, including donors, can invest funding into promoting "success," however it is defined at the local level: (1) by actively engaging with and employing the INEE Minimum Standards for Education in Emergencies, Chronic Crises and Early Reconstruction (MS) as a framework before, during and after emergencies; (2) by promoting and participating in the education cluster as the most effective mode of coordination and support in an emergency response; and (3) by consistently monitoring and evaluating both the hardware and software aspects of education in emergencies preparedness and response in order to ensure the highest quality reaction for each and every emergency. These elements are essential to successful responses and to progress being made in education in emergencies, whether in natural-disaster or conflict/post-conflict settings.

The MS were developed because international humanitarian responses were neglecting education. More than 2,250 individuals representing more than fifty countries participated in the global consultative MS process. The standards reflect global consensus on good practices and lessons learned across the fields of education and protection in emergencies and postcrisis situations as well as the right to education as articulated in the Convention on the Rights of the Child. The MS provide a common language accessible to a variety of stakeholders and promote the continuation of holistic education preparedness and response. Six essential areas are covered by the MS:

- Community participation
- Analysis
- Access and learning environment
- Teaching and learning
- Teachers and other education personnel
- Education policy and coordination

The MS standards, indicators, and guidance notes offer a common framework to the international community for providing protection and coordination for safe access to education at the start of an emergency while simultaneously laying a solid foundation for holistic, sustainable and quality education through reconstruction and development. By utilizing the standards *before a disaster*, the MS also serve as a preparedness tool with which to build the capacity and resilience of education systems. The SPHERE Project (launched in 1997 by a group of humanitarian NGOs and the Red Cross and Red Crescent Movement to increase effective, high-quality response to emergencies) in early 2009 adopted the MS as a companion document firmly placing education (in emergencies) as a first-tier response sector, alongside health, water, and sanitation and shelter.

The MS are a highly effective tool when used in conjunction with the cluster system. Co-led by UNICEF and the Save the Children Alliance, the education cluster makes sure that humanitarian personnel are equipped, prepared, coordinated, and able to respond effectively to emergency situations of all kinds. The education cluster works closely with other clusters, particularly with protection and shelter, in order to provide a holistic, needs-based response. Cluster members work together throughout all phases of an emergency to identify gaps and overlaps and to utilize funding in an efficient and effective manner, the goal being that cluster activities are directly linked into national-level education sector plans.

Emergencies as Windows of Opportunity

Looking at emergencies as windows of opportunity can help us to better understand success and to promote increased donor investment. After the devastating Pakistan earthquake (2005), the education cluster, through its large and active membership, was able to bring children back to school as well as to bring approximately 38,000 children (almost half of whom were girls) to school for the first time. Emergencies often allow the global education community to work at an accelerated pace in challenging environments where innovative strategies and tools can showcase sustainable results more quickly than in development settings. One education cluster meeting in Islamabad during the late emergency phase in 2006 involved a comprehensive debate and discussion around the pros and cons of using mules versus helicopters for transporting school supplies into high-altitude, remote mountain villages. Without this active group participation and continuous, real-time sharing of experience, the cluster would have been unable to respond as successfully as it did in its cost-effective transport of critical supplies, including school tents, School in a Box kits, and recreational materials to some of the worst affected and most vulnerable school communities. During the recovery and rebuilding phase the cluster worked to identify good-quality building contractors and to coordinate the reconstruction process. This concentrated effort on hardware and infrastructure led to the development of related creative activities such as using rubble to build transitional shelters (temporary classrooms with life spans of up to fifteen years), creating community-based tent maintenance guidelines and recycling education supplies during other emergency responses (for example, school tents and extra tent poles were sent to Balochistan and Sindh provinces to create temporary learning spaces [TLS] during the 2007 floods response).

In the Gaza emergency in early 2009, education cluster members, supported by donors, worked with early childhood providers to ensure that young learners received quality pre-school education in conditions where access to supplies had become a major issue. Cluster members used the MS as their guiding framework, encouraging local community participation and working together to identify gaps, provide quality programming, and advocate for improved teaching and learning conditions in a very challenging and highly volatile environment.[14]

In many post-conflict/protracted emergency settings, alternative (often accelerated) learning programs (ALPs) allow children and youth, many who have missed out on weeks, months, and even years of education, a second chance opportunity at learning. When peace and stability finally resume, many older children and youth find it difficult to reenter the formal education system. ALPs in countries such as Nepal, Afghanistan, and Uganda have successfully supported students of all ages to return to school or go to school for the first time. ALPs often condense existing national curricula to capture only the most essential elements and competencies, thus significantly shortening the period of study (per grade year) allowing learners to progress through the system in an accelerated manner.[15] In some cases, older students choose to study vocational skills versus the traditional academic stream curriculum. Donors may consider increasing their support to both ALPs and vocational training initiatives in order to help young people successfully reintegrate into lifelong learning and to becoming responsible global citizens.

Teachers: Key Agents of Change in Emergency Contexts

In order to contribute to successful educational interventions during emergency responses governments must focus on teachers, the key agents of change. According to UNESCO, in the coming decade, the world faces a shortfall of 18 million primary-school teachers. The areas that most need education personnel are countries affected by emergencies and disasters. Teachers are among the most valuable frontline responders in emergencies and strong agents of positive behavior change. By investing in teachers, particularly females, organizations can ensure a good return on their investments in education in emergencies.

For a teacher, defining "success" in both emergency and non-emergency contexts means being able to help children gain access to quality education and to ensuring that meaningful learning actually takes place. Without qualified teachers, a school is just not a school. A powerful example of successful investment in teachers comes from Maira Camp, which was established in post-earthquake North West Frontier Province, Pakistan, and, at its height in 2006 accommodated approximately 20,000 internally displaced persons (IDPs). The primary schoolgoing population of the camp was about 6,000, and an estimated 50 percent of children were girls. The education cluster's objective was to provide access to quality primary school for all children, with a particular focus on girls. In this part of Pakistan, it is not culturally acceptable for girls to attend school, unless their teacher is female.

After the earthquake, there were no suitably qualified local women available who could be recruited as teachers for the camp school. UNICEF and partners in the education cluster established close ties with district education authorities who recruited seven women teachers from Mansehra, a conservative town two hours south of the camp with a far higher rate of female literacy and enrollment of girls at school and thus a better supply of female teachers. These young women were recruited only one month after the earthquake. A number of incentives were offered including competitive salaries (based on individual qualifications and experience) and tents for living in the camp. All the women were related. An older man, closely related to the women, accompanied them to the camp, where he acted as their chaperone and as a paid guard at the school. Camp security was ensured by the presence of the Pakistan army, thus the safety of the women teachers was guaranteed.

The presence, motivation, and professionalism of these young female teachers played a pivotal role in the success of the school in attracting and retaining thousands of girls living in the camp. School visits and classroom observations by UNICEF education officers and partner organization staff confirmed that girls were learning to read, write, and count, to speak in Urdu and English, and, perhaps most importantly, to begin the healing process and regain a sense of normality. It was clear that school was fun. For most students in this remote camp, this was the first time they had been to school. Their female teachers were role models for the camp population, illustrating what was possible in terms of social participation for women, participation based on the acquisition of education, an alien concept pre-earthquake in this remote and highly conservative part of Pakistan. This dynamic example shows how authorities seized the opportunity provided by a massive natural disaster and utilized teachers as key agents of positive change.

The motivation of teachers, principals, and other frontline education personnel during an emergency response is essential. In many emergencies education staff are directly and personally affected by events. In Sri Lanka, for example, many senior education personnel and teachers were IDPs themselves and yet were expected to report each day to temporary learning spaces and to teach children with varying academic and psychosocial needs in extremely challenging circumstances such as overcrowded temporary classrooms, limited resources, and traumatized students and colleagues. Donors must work closely in a country with the education cluster at all levels, national and in the field, to support particular education personnel needs. Costs associated with ensuring sustained motivation of frontline teaching and administrative staff, training costs of volunteer teachers and teacher training on context-related, needs-based topics such as teaching in a TLS require constant donor support. In many emergencies, costs of teachers' kits, monthly training stipends and incentives for teachers (for instance, opportunities for professional development and support, teacher clothing such as saris for female teachers), food, and safe and reliable transportation to and from school are also in demand and require adequate funding. Investing in teachers and their needs during emergencies is critical to effective response and to ensuring substantial returns on donor investment.

Lessons Learned: The Value of Investing in Education in Emergencies

Although education in emergencies is still an emerging field of academic study, valuable lessons learned *do* exist. The problem is that most of the individuals learning and applying the lessons are so busy working tirelessly in the field that they often do not have time to record what they have learned in terms of best practices. The limited or lack of funding in many emergencies results in funds being channeled into the response itself and not into analyzing the quality of the response. Beyond writing the obligatory donor reports, in most organizations there are no staff members specifically tasked with carefully documenting the valuable knowledge and lessons learned from the response experience. It is most effective to capture information in real-time as events are unfolding before staff forget their often innovative solutions. As time passes, a critical literature base will organically and gradually develop. Donors can contribute to the advancement of the field of education in emergencies by earmarking funds to capture valuable lessons learned through the formal documentation of responses.

One good practice and a lesson learned from numerous emergencies is the continued and regular use of the MS from the pre-emergency preparatory (including disaster-risk-reduction) phase all the way through to the return-to-development phase. For even the novice emergency responder the MS provide a manageable framework for successfully responding in an emergency. In the Gaza crisis (2009), for example, the MS rapid needs assessment was quickly adapted by education cluster members in Gaza and the West Bank, with additional inputs from experts in various international locations, to incorporate the particular needs of this complex emergency. Diagrams of school buildings from the Shelter Cluster were included to help data collectors visually describe the damage level of individual schools.

Another valuable lesson learned is that the link between experts from countries of the South and those of the North must be strengthened. There is no blueprint for education in emergencies response. What works in one context may need to be modified in other seemingly similar contexts. It is therefore essential that the international community work to train national level experts and develop their skills as potential MS trainers, education cluster coordinators, and high-quality, frontline responders. The MS require local interpretation and local innovation for effective response. Involving children, youth, and adults with a focus on making education work better for everyone in the community is also essential.[16] In post-conflict Sri Lanka, for example, the education cluster (made up of Sri Lankan and international education experts and stakeholders) worked closely with IDP learning communities to ensure that the varied needs of students, parents, and teachers were being addressed during each phase of the response. During the acute phase, many TLS required school furniture and basic teaching and learning resources. Teachers asked for support in coping with teaching in challenging TLS environments and many students required psychosocial support. Over time, increased community participation became more critical as the complex resettlement process unfolded. A combination of a national understanding of the complexities of this particular post-conflict situation coupled with international expertise in education in emergencies provided a rich environment for innovative, flexible, and results-based response.

Perhaps one of the most potent lessons learned to date is the fact that in the majority of emergency responses, things take time. In general, emergency responses are associated with immediate action and, across sectors, seizing windows of opportunity for innovative and accelerated programmatic response. In fact, evidence from many recent emergencies, including responses to the tsunami and the Pakistan earthquake, indicate that recovery-related activities such as school re-construction, systems support, and capacity development in emergency response and preparedness take considerable time to plan, develop, implement, and evaluate. Pre-emergency, cross-sectoral contingency planning, disaster-risk-reduction, and conflict-risk-reduction measures must be developed well before an emergency happens in order to achieve high returns on donor investment as well as high-impact results.

Recommendations to Donors

Over the past decade, tremendous progress has been made in the emerging field of education in emergencies. However, as climate change, political conflicts, and large-scale movements of human beings increase in intensity, more attention and donor funding will be required for high-quality and sustainable education in emergency response.

In order for organizations of all types to meet their obligations on the right to education in emergencies, there needs to be a focus on education preparedness and response, policy, and funding. National measures must recognize and ensure children's right to education in emergencies and education in emergency preparedness plans should be an integral part of general education plans, including structures for continuity at all levels—from early childhood education through to secondary and tertiary education—and through different phases of an emergency, including specific strategies to ensure safe access for groups that are often discriminated against, such as girls, indigenous people, and disabled children. Emergencies impact particularly severely on people with disabilities. When authorities seize this "window of opportunity" to provide gender-responsive, inclusive education, they can bring about long-term and sustainable changes in educational systems and begin to more equitably share opportunities between women, girls, boys, and men.

The holistic reconstruction of education systems in emergencies, which give a lifeline to children and communities, rests on sufficient funding. Donor governments and the international community should increase long-term predictable aid for education in emergencies and include education as part of humanitarian policy and response in *each and every emergency*. Donors should increase the allocation of education aid in humanitarian crises in line with needs and commit themselves to supporting the Global Education Cluster to ensure that it is adequately funded. Donors can work with education cluster members to develop risk-management capabilities. This is especially important in the area of school construction, where safe, child-friendly schools are those built in compliance with international building codes and standards for disaster resilience while using locally available materials. UN member states must develop plans to replace and retrofit existing school buildings that are unsafe, minimizing risks from building contents and nonstructural building elements. "Building Back Better" using child-friendly methodology, and not just building back to pre-emergency levels, should be the overriding goal.

Donors may consider shifting away from the current emphasis in emergency reporting on measuring quantifiable, but often inaccurate, figures (as figures are subject to constant change) on school enrollment and dropout rates, for example, and greater use of qualitative methodologies that will make it possible to determine the degree and quality of critical software interventions such as psychosocial care during emergencies, from the early acute phase through to development. Teachers, students, and community members can offer guidance on school community needs and how interventions (and success) can be effectively measured at the school level. Donors must listen to children, adolescents, their teachers, parents, and education personnel about what they want and expect from education. Learning is the overriding goal.

Successful examples of innovative, cost-effective models of education in emergencies from varying contexts should be documented and best practices widely shared. The global education community may seek mechanisms to ensure greater and more effective non-governmental organization involvement in the Inter-Agency Standing Committee with a view to improving the coordination of the humanitarian response in the area of education. Save the Children, active in many emergency-affected countries, has an important role to play as the international non-governmental organization jointly lead the education cluster along with UNICEF.[17]

Donors should insist upon and support an intersectoral approach to education in emergencies. Schools can act as a physical point of convergence and as an entry point for providing life-saving and life-enhancing education, protection, water and sanitation, nutrition, and other healthcare initiatives leading to high-impact programming and positive behavior change, not just for students but also for entire communities. Regular and consistent donor support is also required with advocacy issues and with critical policy dialogue input at the national and international level. The combination of both donors and frontline education-sector organizations working together is necessary to move things forward at the highest policy levels.[18]

Donor flexibility and adherence to the Paris Declaration is essential. When organizations on the front lines of an emergency originally apply for funds often the total parameters of the situation, and therefore needs, are not completely known. Donors must offer flexibility in allowing organizations to adapt funding proposals with room for movement/adaptation between activities and associated budget lines in order to meet the changing needs of education sector stakeholders at school, state, and national levels. The case of Sri Lanka provides a powerful example. As the IDPs fled to government-controlled areas of northern Sri Lanka in early 2009, the education cluster worked hard to raise funds to support the building of TLS. These structures, made locally from metal sheeting for walls, pipes for scaffolding, and Cajun (palm) leaves for roofing, were essential in providing safe learning spaces for children living and learning in the camps and transit sites. But as the complex resettlement process unfolded, the needs of IDP children, school communities, and the education sector were ever changing. Increasing focus was directed to continued hardware support, such as TLS maintenance and transportation of TLS to resettlement sites, the continuous replenishment of teaching and learning kits and school furniture, as well as developing critical software needs such as alternative education programs and continued and increased psychosocial support to students, teachers, principals, and education personnel, themselves IDPs working and living in difficult conditions.

Through the education cluster, donors and education personnel need to continuously work together to most effectively utilize organizational area(s) of comparative advantage as a global community of practitioners to share knowledge, lessons learned and innovative ideas of how to provide the highest quality of education in increasingly complex emergency settings. UNICEF is, for example, the lead agency in educational supply provision during emergencies, whereas Save the Children has a comparative advantage and strong track record in Early Childhood Education in Emergencies. In South Asia in mid-2009, INEE members working in Sri Lanka and Pakistan requested support and experience from the wider INEE community as they tried to address the enormous challenges of providing high-quality psychosocial support to students, teachers, and other educational personnel in both difficult contexts.[19]

The Way Ahead

Education is critical for all children, but it is especially urgent for children affected by emergencies. It is about recognizing and meeting basic human rights and human dignity. In addition to being a right clearly articulated in numerous international treaties and declarations, education is an enabling right. Gaining and utilizing the knowledge and skills that basic education affords permits young people to exercise other fundamental rights. Education is also prioritized by communities affected by emergencies, as it gives children the skills they need to pursue peace and the development of their families, communities and countries. Finally, education in emergencies offers a window of opportunity to "build back better" and to work with governments and communities for social transformation by creating programs which allow often excluded groups, like pre–school-aged children, girls, adolescents, and disabled children, to attend school, improving access and quality of education.

The donor community has a critical role to play in emergency response. Donors must continue to analyze the return on their investments in education in emergencies while simultaneously thinking outside of the box and taking calculated risks in their support to the provision of quality education in some of the most challenging and complex emergency settings around the world. Education in emergencies can change lives. Donors have the power and the resources, both financial and human (technical assistance), to contribute to this change.

Education as a Means of Conflict Resolution

PRESIDENT PIERRE NKURUNZIZA

I am proud to announce that the war in Burundi is over! We are emerging from decades of conflict that has been cyclical through recent history, recently in a twelve-year civil war. The last rebel group has now put down its weapons and has turned into a political party, and its leaders and militants are integrated in government structures. Our government is seeking to develop the dividend of peace, with education as a cornerstone. As President, I am committed to this approach, as demonstrated in the concrete investments toward realizing free basic education for all children. But given our limited resources, this alone is not enough. We welcome the political will of the UN General Assembly to assist with encouragement of increased financing from donors. However, I would urge donors to take into account some key factors in building upon an education strategy in Burundi.

The impact of armed conflict on education is overwhelmingly negative. Schools and infrastructure are damaged or destroyed, government spending on education decreased, and students and teachers killed, injured, displaced, or recruited into fighting forces. But the relationship between education and conflict is twofold: education can play, and has played, a role in the outbreak of conflict through unequal educational access and through negative teaching in the curriculum and in the classroom. The case of Burundi has demonstrated these negative effects, but, at the same time, it is important to realize the important positive role that education can play in post-conflict and transition situations by managing conflict, preventing violence and putting a country firmly back on the road to development.

In the aftermath of conflict, education can be an important tool of conflict management. Educational provision is a key peace dividend and can be an important incentive to disarm. Improved educational provision can send an important signal from the government that the state is committed to the well-being of its citizens.

Moreover, improved educational provisions and curricular content can be a critical element of rectifying long-standing group inequalities and of delegitimating violence as a tool of conflict resolution. Improved educational provision and teaching can contribute to the long-term peaceful management of relationships between groups in society, thus reducing the risks of conflict erupting again in the future.

In my contribution, I will discuss the potential role of education in conflict and post-conflict situations and discuss how this has manifested itself in Burundi and how my Government sees education as a determinant factor for the development of our country. I will further analyze some considerations for donor support in the education sector, as they relate to Burundi and other so-called "fragile states." This includes highlighting the importance of recognizing local context as the main reference, dialogue with all stakeholders, attention to the model laid out in the Paris and Accra Declarations, and caution in proposing only "transition measures" when there is momentum for a long-term sustainable education strategy.

Dealing with the Past

Exclusion has been a root cause of much of the armed conflict that has affected Burundi since independence in the early 1960s. In some part this exclusion is a legacy of the colonial past, which fostered an acute consciousness of ethnicity, and in other ways it has been a tool of political manipulation. In no way has this been more evident than in the area of education.

In an economy which has been heavily dependent on peasant agriculture, education has been the only real means of advancement in society, most commonly through state employment. Access to education has been a key determinant as to which groups are excluded, and which groups have access to higher social and economic status. Those in the excluded groups have often been driven to violence by this inequity. A 1999 report on poverty in Burundi, jointly produced by the Ministry for Planning and UNDP, states: "the non-access to education and training constitutes a factor of exclusion from information, and may be the principal source of other forms of exclusion."

In Burundi, schooling has been used as a weapon. Those who were educated were the first ones to be killed in the events of 1972. Later on, education was used for exclusion, with one ethnic group systematically refused access to schooling. Education was seen as wealth. This imbalance in education became a major source of conflict between ethnic groups, between rich and poor, between provinces and between urban and rural zones. Our government therefore stresses the importance of education for all, in order to fight against the fragmentation of our society. All should have equal access to education, as an equal right to develop one's capacities: the same rights for all, which is a factor of reconciliation.

The Arusha Peace Agreement of 2000 was a watershed in seeking to bring a fair and equitable distribution of power and resources to Burundi, as a means of addressing this root cause of violent conflict. The first step in addressing any conflict is to recognize its cause, which Arusha clearly did in highlighting "a discriminatory system which did not offer equal educational access to all Burundian youths from all ethnic groups" as one of the "the causes of the violence and security in Burundi." The Agreement went on to offer the framework through which to change that past reality in outlining "principles and measures relating to education," including:

11. Equitable regional distribution of school buildings, equipment and textbooks throughout the national territory, in such a way as to benefit girls and boys equally.
12. Deliberate promotion of compulsory primary education that ensures gender parity through joint financial support from the State and the communes.
13. Transparency and fairness in non-competitive and competitive examinations.

14. Restoration of the rights of girls and boys whose education has been interrupted as a result of the Burundi conflict, or of exclusion, by effectively reintegrating them into the school system and later into working life.

The way forward is laid out concretely in the Arusha Agreement when it states,

> The Government shall endeavour to correct the imbalances in distribution of the country's limited resources and to embark on the path of sustainable growth with equity. It shall set itself the following principal objectives . . .
>
> (b) Providing all children with primary and secondary education at least to the age of 16

The "imbalances" referred to in the Arusha Agreement are implicitly a call to effect balance, which is difficult to achieve with numerous competing interests. Not only is the consciousness of ethnicity a challenge to making certain that all have equity in access to education, but provincial interests also come into play. It is critical that changes that benefit one group are not perceived as threatening to that of another. Therefore, reducing the educational resources to the formerly more favored provinces is not an option if we want to reduce the risk of a backlash.

The Government of Burundi is committed to the principles of Arusha as a guide toward moving forward, in a balanced and concrete way, in developing our education sector, while keeping in mind the realities and context of the situation in our country.

Looking Toward the Future

In addition to being a cause of conflict as well as a tool of conflict management, access to education is almost universally regarded as a key peace dividend in countries affected by conflict. Around the world, one in three children, at least 40 million, is out of school in conflict-afflicted areas, leaving them vulnerable to violence and facing poverty and instability.

Burundi is proud to be at the forefront of a momentum toward achieving Millennium Development Goal Two (MDG) of ensuring that, "by 2015, children everywhere, boys and girls alike, will be able to complete a full course of primary schooling." In fact, Burundi proposes to exceed reaching MDG Two of providing basic education for all children by 2010, in advance of the 2015 goal. The global MDG Report of 2009 indicates major accomplishments were made in education throughout the developing world, where enrollment in primary education reached 88 percent in 2007, up from 83 percent in 2000. Most of this progress has been in countries that had been most lagging behind, including in sub-Saharan Africa where enrollment increased by 15 percent from 2000 to 2007.

In Burundi's case, in 2005 I took a decision on free basic education for all children. The results were that in 2005, 60 percent of the children between six and twelve years old went to school, whereas today the net enrollment is 85 percent in the course of only a few years. More investments were made, both by government and by our development partners: During the last three years we built approximately 7,000 new classrooms and last year alone distributed 50,000 school benches, with assistance from Belgium, as well as notebooks to all children; 184 secondary schools and 150 primary schools were constructed in 2008 through government mobilization of the population during community work days.

Particularly, I wish to stress the importance of engaging the entire population in our common future. First, the populace shows a strong wish to invest with their own hands in the future. It is highly remarkable that thousands of women and men on our hills, each Saturday morning again and again, engage with their bare hands in building schools without being paid. This shows that my country is fully aware of the importance of education. It shows that Burundians have a great sense of patriotism. Red is one of the three colors of our national flag and symbolizes this love for one's country.

Spending credits on education were increased by 50 percent of the Burundian national budget in 2009 and now has 23 percent of its recurrent budget spent in the education sector, well above the average in similar countries. A recent census of the Burundian civil service counted a total of 57,000 civil servants, of which 41,000 are in the education sector.

In addition to increasing the number of new schools, we put a lot of effort into encouraging parents to send their children to school. Parents are aware of the benefits of schooling, but may be reluctant to enroll their children, which implies an investment that they may find difficult to bear. It is my aim to lift this burden from them.

There used to be direct costs through school fees, which acted as a major barrier to education. In 2005 I abolished all school fees for primary education, as mentioned before. This has had a major effect on school enrollment, accounting for its increase from 2005 to today. We are consolidating this achievement by further reducing the direct costs and the opportunity costs for parents. We are expanding the school feeding program, which is lightening the financial burden for parents and giving them an extra incentive to enroll their children. We are also working on reducing the repetition rate, which currently stands at an unacceptable 36.6 percent in 2007–2008. The aim is to have a repetition rate of 10 percent in 2015.

A new Law on Education will also introduce the principle of compulsory primary education. By doing this, Burundi will join the international community in recognizing education as a universal right, which the State must enforce and support.

Moreover, I will invest much more heavily in the quality of education. If parents feel that their children are not getting the basic knowledge and skills that they expect from primary education, they will not see it as a worthwhile investment on their part. State investment in the quality of education will target the major inputs necessary: increase in the availability of school textbooks, modernization of the school curriculum, recruitment and management of teachers and improved organization in the management of effective teaching hours.

At the same time, the Ministry of Education is spearheading initiatives to integrate conflict prevention into the school curriculum. Starting in 2009, the peace education syllabus will include citizenship and life-skills programs that incorporate peace-building skills. Beginning only four years after the end of the conflict, Burundi is ensuring that peace will be at the foundation of the curriculum at the outset of our rebuilding of the education system.

Reality, and a Rallying Cry for Increased Education Aid in Conflict-Affected Countries

There remains much to be done. We will need to focus on better strategic planning and budgeting; to pay attention to secondary education; to continue building the capacity of teachers; and further raise the national budget for education. Moreover, my Government is engaged to give more attention to vocational and professional training. We have already started to build a certain number of schools to this effect. One of our main objectives is to transform our demographic growth, always considered as a problem, into a real advantage in the development and competitiveness of Burundi.

Although our dedication to this program of action in education is palpable and does indeed bring hope, the comparative statistics are telling that there is still a significant challenge ahead, which will require the assistance of partners.

Consider the educational budget of Luxembourg, spending as much as $27,000 per child per annum on education, while Burundi's national budget as a whole (approximately $500 million) allows only $23 per child per annum. Even with such a large percentage of our budget going to the education sector, some children still have to sit in an unfitting classroom in a class of more than eighty pupils, often without any paper or pencil with which to write. In addition, Burundi is dealing with an influx of refugees returning from Tanzania and the Democratic Republic of Congo that reached more than 94,000 in 2008 alone. The reality may be daunting, but it is also a rallying cry for prioritization of education aid in conflict-affected countries.

Some Advice for Donors

In order to advance the debate concerning the interaction between donor and recipient states, we must move beyond pejorative terminology and view the relationship differently. For instance, the term "fragile states" should be used with caution. Although this is perhaps only a matter of semantics among development practitioners, who now propose "countries in a fragile situation" as an alternative, it still does not remove the sense of paternalism that is attached to these words.

This phrase, apart from the emotional implications it holds, also has concrete financial and political consequences. It creates a negative image in the eyes of foreign investors who are so badly needed to return the country to a development footing.

There is also an adverse effect on the implementation approach taken by donors, and often results in short-term, transitional initiatives to the detriment of a long-term development of the education sector that can be sustained by the government itself into the future. The term "fragile states" needs to be replaced to reflect that sense of hope in the future that is at the very heart of education. Building constructive partnerships based on mutual respect, not on paternalism, should be instilled in that language.

It is from the above premises that the donor community should calibrate its assistance to the education sector in Burundi. The strong call of the UN General Assembly for increased financing, as well as a growing awareness for the need of better coordination mechanisms is commendable. Burundi is not opting for political or economic isolating, because, as our ancestors said, "A man without friends will not be able to cultivate by June" (*Inyakamwe ntirimira impeshi*). Nevertheless, there are a number of considerations and propositions that the donor community should also take into account.

Local Context

Always consider the local context as the main reference, and do not make a single blueprint for all countries. This means that the cluster approach, as advocated in the UN System, is not always the preferable solution. In Burundi, we have sector working groups—bringing the development partners and the Government together—and wish humanitarian actors to align their interventions to existing, instead of parallel, coordination mechanisms.

More fundamentally, it seems that we have lost the link between the needs of our society and our educational system. We have lost the pertinence of our schooling system. Are the skills that we teach our children really the skills they need? Are our teaching methods anchored in Burundian values and traditional culture, or are they copied from the Western world? In my country, the children once learned from their surrounding family, and the people who lived on the same hill, from an oral tradition. We used stories and songs to teach. There is always a need for partnerships that can find an intelligent mix of skills, some as a contribution from the Western world and some from the local culture.

I would again advise heeding the cultural principles laid out in the Arusha Agreement which calls for: "Education of the population, particularly of youth, in positive traditional cultural values such as solidarity, social cooperation, forgiveness and mutual tolerance, Ibanga (discretion and sense of responsibility), Ubupfasoni (respect for others and for oneself) and Ubuntu (humanism and character)."

Sustainable Long-term Development and Dialogue with All Stakeholders

In countries like Burundi the distinction between development and humanitarian interventions in the education sector is very thin. We therefore wish a continuous dialogue with all our international partners, bilateral and multilateral donors, as well as international NGOs. We finalized a strategic plan for the education sector, which I expect all our partners to follow and help implement. The therefore need to work hand in hand with our three Ministries of Education and build the capacity that will allow Burundi to stand on its own two feet.

International support for our burgeoning education strategy is solid and growing. Each international partner is helping, through its own added-value; ranging from project help to support for institutional capacity building and direct financing mechanisms in support of Burundi's State budgeting process and expenditure for education.

I am encouraged to see partners beginning to invest in our national systems, notably through general budget support. I recognize that our systems are still relatively weak in some areas, and I therefore see this international commitment as a mark of trust and confidence in our capacity and commitment to improve our systems.

Several donors are providing direct financial support to our education system through a basket fund, which may at some point evolve toward full sector budget support. This is an important step in the further alignment of donors, not only on our strategic plans for education but also on our systems. The basket fund, nominally set up in 2008, is expected to be fully functional by 2010 and provide significant additional funding for the education sector.

Stick with Existing Models

We propose thus not to reinvent the wheel concerning harmonization and coordination mechanisms. The models as laid out in the Paris and Accra Declarations already exist. I would strongly urge the headquarters of our donors to improve communication with their offices in the field to better implement these declarations on aid efficiency, instead of continuing to produce small, isolated projects. New and innovative financing mechanisms are required—and are well known throughout the world—that allow countries emerging from conflict to build their own national capacity to manage the education sector, rather than implementing transitional projects through parallel structures, such as international NGOs or UN agencies. Capacity can be best built by doing.

I would also advise caution in proposing "transition measures." Measures such as interim sector plans often do not meet the goal of building the necessary capacity for managing more long-term plans. They may also risk prolonging the transition, or "limbo phase," absorbing much of the capacity from national institutions to coordinate, develop and manage these processes, while remaining too "short-term" to deliver sustainable results. In Burundi, such measures have almost pushed us to have a transitional education strategy, whereas we were in a hurry to develop a full-fledged strategy. Transition measures could keep the country in the mind frame of the short-term and "emergencies," whereas we want to develop our strategies for the long term.

Conclusion

Education is high on our Government's agenda as is clearly shown in all our planning documents. It is placed at the forefront in our PRSP (2006–2010) and will also be among the top priorities in our second PRSP (2010–2014). Furthermore, in our long-term strategy, Vision 2025, the education sector as a crucial means to development is greatly stressed.

I am therefore proud that Burundi's Minister for Education gave the inaugural address at the UNESCO International Institute for Educational Planning (IIEP) for the 2009 summer school on "Rebuilding Resilience: Planning Education in 'Fragile Contexts'" in Paris. And although I remain skeptical about the term "fragile" in this regard, I appreciate the recognition of Burundi's achievements in overcoming the armed conflict of its past and becoming a stronger and prouder nation through education.

I am passionate about this topic of education in conflict-affected countries and value the opportunity to represent the voice of governments that are dealing with this situation in their own countries. I seek engagement with partners, both locally and in the international community, to promote understanding in this respect.

Having been an educator myself, I recognize the value and central role that education plays in the formation of young minds, and by extension its role as a determinant of the future path of my country in a direction of peace and prosperity. For it is our children who will make armed conflict a thing of the past in Burundi, and achievements in development the driver of our future.

6 | Hear Our Voices

Experiences of Conflict-Affected Children

ZLATA FILIPOVIĆ

This book contains contributions by many distinguished authors who bring their expansive knowledge, research, and experience to bear on the issue of education in emergencies. My aim is to offer another viewpoint, that of children and young people like myself who have experienced conflict and who lost and (in some lucky cases) regained their education in this particular kind of emergency. I will refer to my experience, that of my friends and colleagues, my fellow members of the Network of Young People Affected by Conflict (NYPAW), and my research into young people's experiences during the time of war. My desire is to bring a child's reality to the issue of education in emergencies and in doing so, I hope, inform further research, policy, and humanitarian responses. Here are stories of real children and young people, their experiences of emergencies, and their experience of education.

All of us referred to in this chapter know what emergencies are—we have felt them on our skin. They crept into our lives, blew them away, sliced them, defragmented them; they stole our innocence, humanity, childhood, families. In all of our cases, conflicts stole one of our basic rights as children and young people: the right to an education. That was the first thing that went when the horrors began. The closure of school was a sign that something was very wrong.

One day in all of our lives, our pens were dropped, notebooks abandoned, benches deserted. Rooms that had been covered with our drawings, lingering with giggles and passed-along notes, became empty. The tremors of fear of being called upon to solve a mathematical puzzle, the excitement of discovering the magic of writing, were gone. Learning how to play, how to pull a pen across paper to leave a permanent mark in this world, to develop one's signature and personality—all were snatched away. Instead, our schools became shelters, places where humanitarian aid was distributed, bombed-out ghost buildings, vandalized spaces, arsenals, demarcations of enemy zones and front lines.

I was born in Sarajevo in 1980. Up until the age of eleven, I thought wars happened to other people. All I ever heard about wars was that they happened far away in history, like World War I and World War II, or somewhere far away geographically, in Lebanon and Africa. Even as the war slowly crept toward my native city, crawling into towns and villages around me, I thought that it could never happen to me. Wars are like terrible diseases, like poverty, like all bad things we know to happen in the world but think will never happen to us. As a natural way of protecting ourselves, hoping and ensuring that we never live them, we prefer not to think about wars, and we delegate them to other people, far away from us.

At the beginning of 1992 I was eleven years old, living a happy and carefree life, looking forward to each day, to going to school, to holidays, to growing up in my city, to working and building a family there. It never crossed my mind that a war would happen to me, turn everything upside down, split my life into two periods: the one before the war and the one since the war began.

On that seismic day I remember trying to write my book report when I heard the first gunshots of my life, sounds that no child, anywhere in the world, should ever hear. I tried hard to concentrate on the assignment, worried what the teacher might say the next day. That was the last book report I did for almost two years of my time in the conflict in Bosnia. My school was bombed and closed, and an enormous rocket hole stood in place of the wall of the literature classroom. I left some nice neatly written essays in a cupboard that was blown apart. I never knew what happened to my teacher. I never saw her again.

Locked inside my house, terrified of the outside world where death could snatch you anytime, I read endlessly, trying to keep myself growing. Then one day, some young women from my neighborhood started a "war school." It was not a real school, and we did not have real classes, but we met occasionally when the days were relatively quiet and we were children again for a moment. It was charmingly called a "summer school," I guess because it took place in the first summer of war. We had different classes we could join—computer science labs (without computers or indeed electricity), dance classes, poetry and writing groups (at least, this is always possible to do).

Some of our teachers in that small room in Sarajevo were eighteen-year-old women whose own education and youth were halted by the war. Others were experienced teachers from before the war, who happened to be living in the few houses surrounding the center where we met for our classes. These women could not watch us children waste away, and they gave us their time and generously shared their imagination, creativity, and knowledge with us. I will never forget them and what they did for us. I can only hope that confronted by similar circumstances, I would be as generous and take on that noble task of a teacher. Then one day our classes were stopped because Eldin, a twelve-year-old boy from my poetry class, was killed by a shell that fell in front of the center. That was the end of summer school.

There were some other attempts to start schooling in the middle of the war, and I remember one day my father coming back from getting humanitarian aid and telling me that real school, not "summer school," would start soon. I gathered my notebooks from before the war, turned them upside down so I could have fresh and clean pages—those blank ones toward the end of the notebook. Since everything in our life was upside-down and back-to-front, why not a notebook too? I was very excited about going to school again, though somewhat frightened of meeting people my age, worried how I would fit in, be perceived, since I rarely saw people of my age during the first year of war.

Classes turned out to be most irregular; instead of times and days being determined by a calendar and watch, they were determined by the intensity of shells falling every day and their proximity to my neighborhood. I went as often as I could, as often as it seemed relatively safe, as often as our teachers could gather and as often as the information that the classes were resumed got to me through the grapevine of the neighbourhood news. Needless to say, this was a most interrupted education schedule, but important to have had nonetheless.

A similar experience of interrupted education is echoed in the words of my friend, Mary Hazboun, a young woman who was growing up in the days of the Second Intifada in Bethlehem. In the summer 2002, she noted in her diary:

> This summer was so special and significant for me. After spending a whole year of massive studying for Tawjeehi [general secondary school examination], I graduated with an honor degree and I got 97% out of 100%. . . . I used to study day and night, during bombing and shelling. Even when there was no electricity, I used to light a candle and study all night. The most difficult and dangerous part during that time was when the students had to go out to their schools and do the tests. It was known that when there was a curfew, no one was supposed to go out. So, we used to stay day and night in front of the TV screen waiting for news regarding the curfew. If they would announce that there was no curfew the next day for couple of hours, the Ministry of Education would announce that the test would be held during those hours. The students would get ready and go do the test during those hours. The most difficult part was when Israeli soldiers used to announce there was no curfew and then change their minds and keep the curfew. So after nights of not sleeping and studying for a very hard and long test, we would end up waiting for the next announcement which could be the next day or the next week. God only knows.

I was lucky in having a chance to leave Bosnia and reenter the world of education, albeit in two new languages—French and English. All I ever wanted during the war, all I wished for, was to go back to school. And then when I finally had my wish granted, I found myself most confused by the people my age suddenly worrying about fashion, what others thought of them, what music they were listening to, and what haircut did not look "stupid." A whole new universe, one that actually belonged to me but was snatched away, now returned, and I was a small lonely planet orbiting in it. It took a long time and many knocks to an A-grade student like myself to fight my way through new languages, knowledge gaps that existed from the war, and the whole resocialization process that confronted me. All of my experiences happened in Europe, and my own experience, despite all its tragedies and problems, was an entirely more privileged one than some others.

My friend Kon Kelei was born in Southern Sudan, approximately in 1980, the same year as me, though he never really had a birth register of his real date of birth. Kon never really had the childhood that I was privileged to have lived. His education started with war itself, when he was taken away, at the age of five, as one of some 27,000 "Lost Boys" of Sudan to a training camp. His primary education started in the camp where he was trained, an education that included arithmetic, some English language, history, and later small amounts of science such as biology. School was not school, not real school—even if it looked like school, emphasizes Kon. The most important aim was to make young soldiers out of these children, for learning how to fight was the main priority of the Sudan People's Liberation Movement and Army (SPLA).

A day in his life in the camp consisted of two hours of schooling in the morning, without books or papers, writing on the ground. Kon jokes today that in this way no one could ever really do their homework, so it was a good excuse. After this short spell of classes in the morning, the books were put down and the rest of the day was taken up with duties in the camp. The only other moment of recreation was half an hour of football or swimming. There was no drawing, painting, playing—no things that make you a child. The only singing that Kon did was of revolutionary songs created by the children themselves.

Kon remembers his first years at school after deserting the SPLA and moving to Khartoum on his own. He was fifteen years old, completely alone, and working odd jobs to put himself through school. He was not aggressive toward his teachers and classmates, but he had huge mistrust toward them all, and having fought himself, he considered them all cowards. What the SPLA had "taught" him, like so many other child soldiers around the world, was that the only way of solving problems was through fighting. A strong and tough man is a man with gun, and no one else can tell him the difference between good and bad; there is no authority. It was a slow revelation that sometimes fights can also be won by one's mind and that you can bring your opponent down to his knees by making a strong argument.

Kon lived in a permanent fear of being found out for what his past life had been, saying, "When you open your mouth, you call death upon yourself." He kept to himself, quiet and observant of other young people and teachers. Slowly, with time, he started recognizing those with the same experiences as him and began forming bonds. Learning to trust his teachers and classmates was a way out of the past experiences and a beginning of a new way to live a life. Education allowed him, as a war child, to gain back a sense of humanity, to open his eyes to others, to become a social being again through his interaction with others. Without this, he says, the effects of war are carried until they explode somewhere along the line and hurt more people.

Education also gave him confidence in himself and his ability to reason, and he discovered his own independence. It suddenly became possible to get the things he wanted, this time without a gun or hurting someone else, and to achieve some of his goals. This self-confidence eventually motivated him to move to the Netherlands, where he is studying to become a practitioner of international law, the same law that states that he had a right to a childhood and an education.

Kon believes that school is not only a place where one sits and learns, but also the place where you meet friends, play openly without fear, learn how to draw flowers and not bullets, a place where you learn what love and peace are. He runs a foundation today that is rebuilding a school in southern Sudan and helping young people continue their education after primary school. He believes that the only way Africa can escape the cycles of violence is if its people recognize the difference between good and bad governance, if they claim their rights, and if they become active members of civil society. This will happen only with education. Kon stresses that education is the only sustainable solution for Africa. You can ask many "why" questions when you have studied.

Another former child soldier, Ishmael Beah from Sierra Leone, speaks of the potential of school existing during war as a possibility of minimizing the recruitment of children. He believes that he might not have been mobilized had he known of a place that he could go to, somewhere where he could have been safe and protected. These words recall images and reports of many Ugandan boys and girls who became "night commuters," leaving their homes before dark to travel to the city centers, sleeping in churches, hospital lobbies, and bus shelters. Perhaps if there were schools as safe havens, protected zones, they could play a double role of safety and education for many children.

Coming out of the war in Sierra Leone, there were many things that helped Ishmael to recover, the rehabilitation process and a strong family being key. However, that holistic healing was possible because of the access he had to education. Through school he learned to strengthen the purpose of his own humanity again and to reaffirm that he was capable of much more than violence, as he had come to believe in his childhood years. He also believes strongly that education does not have enough value in the eyes of broken, disappointed, traumatized youth, who have been let down by their governments and the international community. It is important that institutions that can change this mindset are rebuilt, so that all young people, like Ishmael, can have the opportunity to regain their education, humanity, and hope in a better future.

The thirst for learning in the midst of most difficult times is well explained by my friend Hoda Thamir Jehad, who is from Nasiriyah, Iraq. Before the war, Hoda was a good student, and she stubbornly continued to hang on to her ambitions, even when the situation in her country was impossibly hard. Sometimes these moments of extreme thirst for life, for learning, for hope, for future are common threads binding so many children and young people of war. Despite everything around us being destructive, we somehow still manage to retain some creativity. Hoda's words below exemplify that in a beautiful and poignant way:

March 23 2003

Today the bombardment was unceasing and very violent, but it was worse in Baghdad and the other governing areas. We do not know what to do. Everyone is at home and no one goes out to work, or to schools, universities, or anywhere in Nasiriyah except for the hospital, which constantly receives the dead and the wounded, where benevolent doctors are sacrificing themselves, giving their souls for the innocent; for them, it's a day like any previous day.

April 4 2003

A day no different from other days, as usual I woke up to the sound of rifles ringing like school bells in my ears; I had to listen to them as there was no escape from the noise

April 7 2003

Today's morning was very normal, and around 10 o'clock in the morning I heard a piece of news regarding schooling. They said that schools will reopen to receive their children and pupils, and the news was confirmed later in the afternoon when an American car arrived, broadcasting in a very loud voice, encouraging pupils to go back to schools, and employees to go back to their work. At that moment I could not contain my happiness; I thought I would die or faint from joy. I started laughing, and running and screaming with joy at top of my voice, and I ran around the house saying: "I will go back to school," while scarcely believing the news. But the important thing is whether my family will allow me to go to school in such extremely disconcerting circumstances; I do not know! Nevertheless, I started calling all my friends to deliver the good news so they might share my happiness with me. But the happiness was soon over because my family refused me, which made me hate this situation in which I live, and the miserable life to which I was born. But there is a glimmer of hope that life will go back to what it was used to be in the past, perhaps even better.

Though her last words bear hope for the future, children such as Hoda, Kon, Ishmael, Mary, and I—every day, all around the world—go into cellars and hiding places, into refugee camps, or into the army. With them goes the future of their countries and of the world. They die; they are maimed, traumatized, broken, these same children who could be future leaders, civil servants, scholars, fathers, mothers, and teachers. Some of us are incredibly lucky to have had the chance to return to our studies, but too many others do not have that opportunity.

All different voices of youth affected by conflict speak of the same idea—schools are where we realize our potential, where we become social beings, where we grow and develop as functioning, contributing, and empathetic human beings of our communities and the world. After conflict, schools are where the dangers of landmines can be disseminated and the prevention of HIV/AIDS can be taught. This is where guns can be exchanged for knowledge and training and where long-term strategies can be developed that interweave peace-building messages with skills and knowledge. While the Millennium Development Goals address the need for primary schooling for all children by 2015, we mustn't forget that more is required beyond primary education so that these children, once they are set on the right path, can reach their full potential. This is not only for themselves but also for the stability, prosperity, and peace of their countries.

If sustainable peace is to be attained, young survivors of conflict firmly believe that education should be an integral part of every peace agreement and strong attention should be given to all education projects in conflict and post-conflict countries. Education begins the long journey for each war-affected child to reclaim their youth and discover their own humanity and their contribution to the human society. It is also an antidote for violence in any society. Education gives young people the ability to use their minds in a positive and constructive manner and thereby enables them to have the capacity for transformation and to build or repair the foundations of their dreams and hopes.

It is also important to mention one more aspect—that of education beyond immediate conflict. How can education be used to sow the seeds of reconciliation and not add to the perpetuity of violence and hatred? An unfortunate example of the latter, close to my heart, is that of the "Two Schools Under One Roof" programs that mushroomed in post-conflict Bosnia and Herzegovina.

Almost fifteen years after the end of the Bosnian war, children are being separated along ethnic and religious lines. The postwar Bosnian phenomenon of "Two Schools Under One Roof" means that children of different ethnic groups use the same school facilities but have no contact with one another, and they follow divergent, ethnic-based curricula. They call their mother tongue a different name; they learn different versions of history; they talk about the conflict that tore their country apart in completely different ways.

In the fifteen years since the Dayton Peace Agreement, thousands of children have been educated in this way. How can there be hope for a sustainable peace if people have grown even farther apart from each other? If the "other" has been distanced as far as possible, sometimes becoming the "enemy"? Learning the lessons of war in a sensitive manner; openly discussing genocide; questioning how to teach history post-conflict, are all issues that deserve careful attention—whether in Rwanda, in Bosnia, or anywhere in the world.

We also need to include in our deliberations children's and young people's voices and views on justice, education, and reconciliation, as well as their fears and hopes. Children are remarkably resilient beings, far beyond what we can imagine, and they deserve to be decision makers in the processes that concern their future. They have a right to express ideas to best assist those who will help them achieve their utmost potential.

I would like to end with a quote from one of the most famous children of war: Anne Frank. As a most talented writer who quietly grew and developed herself in spite of the most terrible circumstances, she is a representative for so many who were killed in wars. Her joy and belief in future are even more poignant as they came to us following the tragic end to her life. I offer her words as a symbol of hope and optimism:

Everyone has inside of him a piece of good news. The good news is that you don't know how great you can be! How much you can love! What you can accomplish! And what your potential is!

7 | Learning from Children

ROBERT COLES

In 1956, I was a resident of the Massachusetts General Hospital in Boston, learning how to become a pediatric psychiatrist. At that time an epidemic of polio broke out—this was before the Salk vaccine had arrived—and a number of youngsters were hospitalized on our wards. Some of those boys and girls were in great trouble—unable to walk or use their arms, or, alas, more desperately, were spending their time in what were then called "iron lungs."

One day a pediatrician teacher at the hospital suggested that I go talk with some of those young patients, and soon enough, an initial visit turned into a longer stretch of bedside visits. Once, as I was talking with a youngster named Patrick, who had lost use of his legs and lay interminably on a bed—even as he, a junior high student and vigorous athlete, kept wondering whether he'd ever be able to get up, go about a normal walking, football playing, school-attending life—I found myself hearing the following words from Pat (as he asked me to call him): "Hey Doc, I'm worried about you—you look tired. Maybe you should take a few hours off, go get a nap." I stood there and was all set to reassure my young patient that I was all right. I said, "I'm okay, okay." Yet, I knew in my heart of hearts that I *was* tired, very much in need of a nap or longer stretch of sleep. I stood there silent for a while, but then got up the good sense to say, "Yes, you're right, doctor!" That last word got a grateful laugh from Pat—who had really become a kind of healer, willing to offer sage words to an all too worn down young doc.

When I told of this exchange, this meeting of minds on a "polio ward," as we sometimes called it, one of my listeners and teachers,

Dr. Carl Binger, a distinguished psychiatrist and psychoanalyst, responded with these words, which I have kept in mind, now, for half a century: "Our patients, now and then, teach us as we try to get them better—and sometimes they even try to help us out: sensing the vulnerability of their doctor, they try to be healers as well as people suffering, their lives in jeopardy." Finally this, a while later: "We learn from those we try to teach, or help deal with disease—and a word of gratitude from us, to a youth like your friend Pat, can go a long way." When I first heard Pat's concern for my obvious exhaustion, I assured him that I was "okay, okay"—but later I did go back and thanked him warmly for his kindly interest in me, and I also told him that I'd gotten a night's sleep. Moreover, I let him know that I would really try to find a few hours of rest, for the sake of my patients as well as myself.

In 1960 I had finished two years of work as an Air Force psychiatrist, doing my duty under the old Doctor's Draft by working at an Air Force hospital in Biloxi, Mississippi. One day, on my way to a meeting in New Orleans, I found myself unable to get into the city. The streets were blocked, and hundreds of people had turned out to protest in anger the court-ordered desegregation of the city's schools. I had my Air Force uniform on, and got out of my car and asked a policeman what was happening. There we were, two men in uniforms, talking about a city engulfed in racial change and protest. The policeman told me what was happening: that a "little colored girl" was trying to get into a school down the block, and as a result people were protesting the order of the Federal Judge J. Skelly Wright that she be allowed to do so.

I was ready to let this matter drop, but my wife, Jane, who taught English and history in high school, felt that I ought to try to meet this girl and find out what was happening in her life. I well remember my wife's words: "You were interested in stress in children who have polio, what about this little girl going through a mob, who in its own way is trying to paralyze her!" Well, encouraged by my wife's way of looking at not only what was happening in front of us but also what had happened in the past in a northern hospital, I decided to try to get to know this girl and find out how she was handling this very stressful life.

I've described Ruby Bridges and her brave struggles in a number of articles and in a book, Children of Crisis: A Study of Courage and Fear. Many a times I've described Ruby's response to the dangers she's faced, the threats by grown men and women that "she oughtta die" and "she very well might die," fighting for what had been declared the law of the land by a federal judge, and eventually by the Supreme Court.

One day as I was getting to know Ruby, learning about how she was doing, all alone in the Frantz School in New Orleans, I decided to become a little more the psychoanalytic psychiatrist, at least explicitly so. I started asking Ruby directly what crossed her mind as she went past that hectoring, and sometimes very dangerously threatening, mob kept at bay only by armed federal marshals. "Ruby, perhaps you could tell me what crosses your mind just as you go by those people?" What was crossing my mind was that there was a good deal of fear and anxiety in her mind, and perhaps my job as a documentarian would truly be to hear from Ruby directly what she thought about the experiences she was having, even as they were going on during her morning and afternoon school life. Ruby hesitated for a while as she pondered what I had asked. And then she said this: "I always think the same thing when I hear those people trying to scare me real bad." She didn't tell me what came to her mind but, of course, I was quite ready and willing to ask her what it was that she always thought. And then this: "I always pray for those people."

At this point silence held fast over the room. Ruby's parents were standing across the room, shaking their heads and nodding. My wife was nodding, too. I'm afraid I have to admit I was not nodding—I was wondering, in the shrink mode, what she must be doing to protect herself psychologically from the severe threats facing her as she tried to pursue an elementary-school education. More silence then, as I dealt with a certain incredulity in my mind. I broke the silence this way: "Ruby, you pray for those people!" Again, silence. And then this from Ruby: "Well, don't you think they need praying for?"

Now I was stunned into my own kind of silence: part of me trying to figure out a traditional psychiatric response to this child's way of thinking about what she was experiencing—all those "psychological defense mechanisms" were crossing my mind. I guess the doctor in me was thinking that Ruby was angry or frightened but couldn't deal with it directly, and it was my job to help her deal with it.

At this point, my wife spoke: "Thank you so much, Ruby, for helping Dr. Coles understand what you're going through—I think you're giving him a good education." At that point I could only stop and think about how much I, a physician and teacher, had to learn from this young New Orleans first-grader. Later, when my wife and I told Anna Freud what we had experienced at that time and place, we heard this: "It took me a long time to learn how much children have to show us and tell us—I would even say, educate us!" For Ms. Freud, that was a rather lengthy response, and one that I think Ruby Bridges in her own way had well earned.

In every discipline, techniques are developed that allow new generations to improve the quality of products and programs. This is as true in education as it is in medicine. Almost every author in this book has emphasized the need for competent teachers if education, especially in emergencies, is to succeed. To achieve that goal, one must build a foundation based on analyses of best practices in previous crises. In this section, I have asked scholars and humanitarian workers to use their own academic studies as well as field experiences to help us both understand the lessons of the past and suggest better ways forward.

8 | An Unexpected Lifeline

GERALD MARTONE

Providing educational services for children is a vital intervention during emergencies, chronic crises, and early phases of reconstruction. The spectrum of activities within the sector of Education in Emergencies are wide ranging and include nonformal education, basic literacy and numeracy, cultural activities and creative expressive outlets, sports and recreation, health education, life skills, peace education, teacher training, support to Community Education Committees, youth leadership, civic development, school rehabilitation, vocational training, and capacity building for host governments.

The primary mandate of relief organizations is often limited to assistance programs that are categorized as lifesaving in scope. These initial activities typically involve the direct provision of food, shelter, water, and other essential services. Physical survival is regarded as a humanitarian imperative and all other concerns are relegated to a subordinate status. The implication is that if precious resources were diverted to less immediate realities, somehow people would be allowed to die.

Yet in the majority of today's protracted refugee situations, people are not dying at unusually high rates. Despite the folklore of our work, these crises are more often not life-or-death situations. Rather, the predominant experience of refugees is a hopeless and purposeless existence: "Improved children survival rates are not very meaningful if children reach their fifth birthdays but are doomed to lives of misery."[1]

The stark reality is that the average length of refugee displacement globally is seventeen years in duration.[2] It is common to find a generation of children raised without any access to education among the world's refugee "warehouses." We must shift our obsession from how people are dying to how people are living.

The Perverse Humiliation of War

According to the WHO, the twentieth century was the most violent period in human history.[3] Unrestrained fighting, ethnic violence, and political oppression still ravage over thirty countries around the world. Contemporary warfare, especially since the end of the Cold War, is a particularly brutal conflagration of terror tactics in which insurgents, mercenaries, and nonprofessional armies—ignorant of or unchastened by the Geneva Conventions and the rules of war—routinely deny the most basic human rights of civilians caught in the fighting.

Torture, child conscription, mutilation, rape, forced labor, and deliberate starvation are among the arsenal of terrorizing strategies of ruthless fighters. Militias, rebel groups, and malign regimes use violent deportation and forced migration of citizen populations as a deliberate military strategy.

The antecedents to a life in exile in contemporary conflict often disrupt the integrity of family units. Families split up as parents seek safety in other countries, join fighting forces, or search for work to support their children. The incapacitating debt that uprooted families incur in some conflict-prone countries such as Afghanistan has led to children being used for collateral on loans, girls being sold to feed their families, boys being bonded to merchants, girls being married at very young ages, and children resorting to sexual bartering or "survival sex."

The effects of conflict on children go beyond physical consequences to include psychological as well as social repercussions:

> Not only are large numbers of children killed and injured, but countless others grow up deprived of their material and emotional needs, including the structures that give meaning to social and cultural life. The entire fabric of their societies—their homes, schools, health systems, and religious institutions—are torn to pieces.[4]

Under the current assumptions and paternalistic attitudes that influence the tradition of aid delivery, educational interventions for children and youth are treated as a lower-tier priority. Education in Emergencies is only just beginning to achieve some stature among the spectrum of vital relief interventions.

Children are an especially vulnerable group of war-affected populations because of their dependence on adults and on their communities for survival:

> Conflict and displacement can present particular threats, such as separation from family, abduction, or recruitment by fighting forces or exposure to targeted violence or landmines. At the same time, pre-existing threats, such as sexual or gender-based violence, labor exploitation, or malnutrition and disease, may increase.[5]

The urgent priorities for life-saving assistance distract us, understandably but regrettably, away from the medium to long-term restoration of livelihoods, economic development, education, and psychosocial well-being. Although only an infinitesimal fraction of refugee crises are short-term, temporary and makeshift interventions with relatively short planning horizons are the rule: "Emergencies are internationally interpreted as occasions for swift action, not as opportunities for critical reflection."[6]

Humanitarian aid is one of the few disciplines where one can actually be criticized for being too focused on long-term outcomes. The crucial intervention of Education in Emergencies struggles for legitimacy among the purveyors of this line of thinking. The current orthodoxy suffers from a hierarchy of assumptions and informal triage that subordinates anything but physical survival.

More than 77 million school-aged children around the world are not in school. Even more astounding is an additional 150 million children who have been forced to drop out of school after less than four years of education due to the pressures of their families' poverty.[7] The world's pledge for universal primary education by the year 2015 is an empty promise.

This problem is even more acute in war-torn countries. According to the Institute of Development Studies, over half of the children and youth who have not completed primary school around the world live in countries affected by armed conflict. Among the many millions of refugee children and adolescents in the world, more than one-third are out of school with no hopes of ever seeing the inside of a classroom.[8]

The impact is even greater on girls. There are more girls around the world who don't attend school than there are girls in all of North America and Europe.[9] In rural Africa, only about 30 percent of girls finish primary school.[10] Yet for refugee children—both boys and girls—only 6 percent are enrolled in secondary education, with even lower numbers among internally displaced children.[11] Write Rebecca Winthrop and Jackie Kirk, "*When children's lives are disrupted by conflict, emergency education becomes a protector and a healer.*"[12]

But People Are Dying

High-profile crises are an irresistible call to action for the disaster-relief industry. "It's an emergency" becomes the mantra as we readily focus our efforts on the transitory emergency needs of populations in distress. We are compelled by the impression that "people are dying" and the notion that "we have no time."

There is abundant empirical evidence that a long-term view of disaster assistance from the outset results in more effective short and long-term outcomes. Yet we continue to react to each new situation as an emergency, invoking the same tired and beleaguered rhetoric: "Can't you see we're saving lives?"

Physical survival is the sine qua non of our work, even when survival in not at stake. Some of the highest-profile crises of the last decade—Bosnia, East Timor, and Kosovo, for example—did not consistently achieve the mortality threshold that qualifies as a complex humanitarian emergency, that is, one death per 10,000 people per day.

Crude mortality rates have become the absolute index of human suffering. The jargon and ritualized traditions of emergency assistance reinforce the myth of incontestably essential survival interventions. In "Questioning the Solution," the authors admit that "clearly, no death statistics entirely reflect the health or quality of life of survivors."[13]

Mortality rates, like the canary in the coalmine, are a trailing indicator—the equivalent of evaluating the public health of a community merely through post-mortem examination and autopsies. Yet mortality rates have become the nonnegotiable rationale for all that we do. It is the humanitarian bottom line.

The urgent, critical, and sometimes rash ethos of emergency assistance is encoded into the humanitarian culture and vernacular. We too readily invoke the emergency alibi in the presence of dire circumstances as an excuse for providing only the most basic and rudimentary care in the name of saving lives.

As populations recoil from frightening abuses and grotesque traumatic experiences, aid projects do little to provide an alternative or even temporary reprieve from their bleak and dreary life in exile. The tedious regimen of camp life offers few distractions from the haunting recollections of their suffering and loss. Large-scale, broad-sector approaches to emergency relief merely emphasize material and physical needs. Said one Ethiopian refugee, a father, "We thank you for helping us, giving us food, shelter, medicines, but the best that you have done for us was to give our children education. Food and other things we will finish but education will always be there wherever we go."[14]

A critical gulf remains unbridged as the complicated emotional needs of people who have survived a brutal and confusing war are overlooked. In the remote central highlands of Afghanistan in the winter of 2001, the World Health Organization reported that "one of the most prevalent reasons patients visit the health center is fear . . . the greatest health problem facing the people . . . is psychological distress."[15] This report was from Hazarajat, the center of Afghanistan's famine-prone hunger belt. "Many [women and girls] suffer the humiliation of having to beg, yet are punished for roaming the streets without male accompaniment."

In *The Selfish Altruist*, Tony Vaux states,

> Aid that simply provides calories for the stomach and water for the throat is a reduction of people to things . . . Concern for the person entails

concern for the whole being, including a person's state of mind, sense of loss and the devaluation of life. . . . It is concern for every aspect of a person including their loss of relatives and way of life, their disability, their love of children, their past and their future.[16]

The monotonous, boring, and uneventful experience of living in a refugee camp is particularly toxic for a child's psychological development. A child's developing mind requires structure, routines, and stimulation for healthy development and the prospect of normality.

Yet in aid work, physiology consistently trumps psychology. The tools of the trade are trucks, warehouses, and sophisticated communication gear. Some relief workers have cynically referred to the habitual commodity-driven aid culture as "truck and chuck" programs administered by "boys with toys." Refugees are treated like cattle where feeding, watering, and population-based interventions are prioritized for the herd.

The field of humanitarian assistance is compartmentalized in such a way that our attention is focused on immediate lifesaving measures and readily quantifiable indices like morbidity and mortality rates.

Tyranny of the Urgent

Aid workers' strategies for coping with human suffering, coupled with the short attention spans of the public and donors alike, drive the development of this short-term thinking. Who can resist the allure of dramatic, immediately tangible, and quick-impact emergency interventions? The emotional pressure borne by aid workers in the face of staggering human suffering is considerable. The cries for help are shrill, and relief workers have an insatiable personal commitment to alleviate human suffering. Yet invoking this life or death hierarchy seems to run counter to our dedication of providing refugees with what is truly the most appropriate intervention.

Economic development, education, and personal dignity are dismissed because "lives must be saved." Longer-term considerations are somehow seen at odds with the frontiersman spirit of disaster relief.

This neglects the harsh reality of life in exile: population displacements due to political upheaval and conflict can last for decades. On the occasions that we are scrutinized for shoddy craftsmanship, we rebuff outsiders. They do not bear witness to the struggles of relief work, and thus have no right to critique those of us who do. As observed in one study of aid workers in crises, "the 'if you haven't been there, then you cannot understand' attitude, if unchecked, can reinforce . . . a fearful, self-referring isolation which puts its members beyond criticism, and their work above analysis."[17]

At times we conspire in this drama through the emotional appeal of our public messages. The portrayal of suffering is presented as short-term, acute crises that are amenable to financial pledges and the charity of Western do-gooders. This perpetuates the action-oriented, anti-intellectual culture of emergency work. Instead of investing in the underlying vulnerabilities of a situation, we focus on shortsighted relief strategies that can inadvertently do more harm than good.[18]

In this culture of urgency, it is no wonder that Education in Emergencies had to clamor for a legitimate place as a stand-alone cluster alongside Health, Nutrition, Water & Sanitation, and other components of relief work within the Inter-Agency Standing Committee's Cluster Approach. Yet this is the most significant reform of the humanitarian system in over thirty years.

By reacting to each new crisis as an emergency, we miss the opportunity for external aid to be catalytic and exponential in rebuilding refugees' lives and livelihoods. This does not have to be the case. By recognizing Education in Emergencies as one of the pillars of humanitarian assistance and respecting the codification of Minimum Standards by the Inter-Agency Network for Education in Emergencies[19] as the professional culmination of our guild, we take the first step toward minimizing the urge to react reflexively with shortsighted imperatives.

Ask Them What They Need

In our more fallible moments, the craftsmanship of aid work has been criticized for conforming to three misguided presumptions: "All aid is good, one size fits all, and we know what's best." This overreaction of paternalism by charities to people in distress seems to imply that the recipients lack adaptive, innovative, and competent survival instincts.

Despite the overwhelming reality of capability and competence among war-affected communities, some relief programs have been described as "the last bastion of the ultra-paternalistic approach to aid and development. It is hard to think of another area where the blinkered nonsense of the 'we know what is best for them' approach survives so unchallenged."[20]

This assumption defies the reality that almost all communities affected by disaster and war organize quickly during a crisis, readily identify representative leadership, invoke indigenous altruism and their own legacies of extending help to their community, and can readily delineate clear needs of what they require to recover.

The presumption that underlies the traditional "needs assessment" is that people are "needy." Rather than presume that refugees and displaced people are passive, helpless, and needy, practitioners of aid should conduct *capacities assessments*. The assessment of a population should include the community's capabilities, assets, and ambitions as part of the evaluation. Aid workers should strive deliberately to involve refugees through the entire planning cycle of a program.

In the conclusions of the *Active Learning Network for Accountability and Performance in Humanitarian Action Annual Review 2002*, participation and capacity building were still found to be problematic:

> Facilitating community participation in planning and decision-making continues to be problematic, no examples of systematic good practice were found in this year's reports. . . . The evaluation team found very little evidence of beneficiary participation in the assessment and program design phases . . . beneficiary participation was largely limited to the "we provide the materials, you do the work" approach.[21]

It appears that despite our devotion and focus on assisting populations in extremis, we are insincere when it comes to asking their opinion and respecting their response. This oversight is brought to light in the countless assessments of displaced populations where refugee leaders specifically identify education and schooling as a priority need for their communities.

Within the maelstrom of destruction and deprivation, refugees instinctively develop their own strategies for their families to survive. Unexpectedly the lifeline they often reach for is education. Some refugees have even traded food in order to afford schooling. Atuu Waonaje was fifteen years old when he grabbed his younger brother by the hand and fled the Democratic Republic of Congo. Waonaje recalls, "To pay for school I had to sell some of the food we received from the WFP, even though it wasn't enough to survive."[22]

The outside observer might note the apparent contradiction: humanitarianism's hierarchy, comprising experts and professional aid workers from wealthy nations, develops policies and approaches that guarantee survival of the body while aid recipients struggle to preserve their minds and the souls of their societies. In Atuu Waonaje's case, he knew the value of education. Eventually Waonaje learned and then taught English. While in secondary school himself, he founded the Center for Youth Development and Adult Education. The internationally recognized Center now has forty-four teachers and conducts classes and workshops in subjects ranging from social issues and health education to occupational skills and basic literacy.

To any experienced aid practitioner it comes as no surprise that education and schooling dominate the conversation when establishing a presence in a camp or settlement. Education is so vital to communities that, even during high-profile emergencies, recipients often identify support for schooling as *the* priority intervention. In many cases the demand by refugee leaders for children's education can exceed requests for food, water, medicine, and even shelter.

During the famine in Afghanistan in the winter of 2001–2002, Western aid groups, whose agenda favored food supply and other commodity distributions, denied village leaders' requests for education. In response, community leaders requested that teachers be categorized among the "most vulnerable" for priority rationing of food parcels. Education was so important to them that they wanted to make sure that teachers did not leave their communities in search for food, wage labor, or other means of sustenance.

In most countries, schoolteachers are respected as authorities and are often sought out as highly regarded sources for advice in their communities. In Afghanistan, when asked what they want to be when they grow up, most children readily respond that they would like to be a teacher, a doctor, or an engineer. An educated person occupies a high place in the totem of most societies, not just in Afghanistan's social hierarchy.

The instinct for people to enshrine education as a community priority is also a notable feature among Chechen displaced families. Many Chechens had abruptly fled their homes in the early summer of 1999 when most children had only sandals on their feet. With the harsh winter approaching, the International Rescue Committee distributed a large consignment of children's boots in several displaced camps.

At a follow-up visit to the camps when snow was on the ground, the IRC staff was perturbed to find children still barefoot or wearing sandals. The families readily produced the children's boots for inspection, but when asked why the boots were still in their original wrappers, the children explained that they were saving their new footwear for the first day of school.

During an assessment mission in Iraq in 2003, the IRC were exploring the feasibility of a small community grant program for displaced communities. Despite the austere and deprived circumstances that many of the displaced people found themselves in, communities consistently identified "building a school" when asked what they would do if they were awarded a $5,000 rehabilitation grant.

In May of 2000, during a survey of displaced camps in the Maluku Islands of Indonesia, leaders of the displaced centers repeatedly requested "schools for our children" as the primary need in their community. Despite the lack of clinics, latrines, wells, and other relief commodities—the traditional preoccupation of aid workers—communities identified education as their number-one priority.

Several relief agencies in West Africa have a longstanding tradition of educational support in displaced and refugee camps throughout the region. In many of these settings, the work of international aid groups merely augments efforts already initiated by displaced communities on their own. In several camps, community members had efficiently formed Parent-Teacher Associations (PTAs). The PTAs identified abandoned buildings and quickly established makeshift schools. These PTAs approached relief organizations for support such as desks and chairs, school supplies, books, teacher stipends, and roofing material. The school system arose spontaneously among the displaced communities, and volunteer teachers were quickly identified among the camp's population.

Sometimes following a large-scale bulk distribution of commodities in refugee camps, it is not surprising to find that recipients immediately sell or trade their cooking pots, itchy wool blankets, and flimsy shelter material to barter for school supplies. They realize that education can alleviate the dreary and hopeless camp settings that imperil children raised with no sense of hope, optimism, or prospects of a better future.

Whom Are We Helping?

In contemplating priorities of intervention within the spectrum of aid projects, it is crucial to scrutinize the age-demographic of the populations that we serve. In most of the developing world, in particular sub-Saharan Africa, as much as half the population is under the age of eighteen. This peculiar skewing of age and the significant prevalence of youth is even more pronounced in forcefully migrated populations. Many adults—men in particular—have been killed, joined fighting forces, or have been forced to separate from their families to seek wage labor or personal safety. A refugee camp is a young place; it is overwhelmingly populated by children. We cannot afford to let this reality elude us in the design of relief programs. That is an inexcusable oversight.

Assumptions, traditions, and ill-informed mindsets undermine the vital position that Education in Emergencies should occupy within the relief guild. Although our profession has become enamored with "evidence-based" interventions, we often miss the very intervention that our beneficiaries are requesting. This has led to a misdiagnosis of their condition and a misinterpretation of the remedies.

A woman who had fled Darfur into Chad in 2004 offered an elegant observation on the importance of schooling in her society. She remarked, "We had to leave behind all of our possessions. The only thing we could bring with us is what we have in our heads, what we have been taught—our education. Education is the only thing that cannot be taken from us."[23]

Considering the precarious and protracted limbo that defines the life of a refugee, this insightful woman is instructing us that one of the most sustainable interventions we might provide is, in fact, education. It is imperative that relief practitioners respect the appraisals by the very recipients of our services; they are not the object of our work but rather the subject. The current configuration of liability in the humanitarian enterprise is an upward accountability to the donors who provide financial support, rather than a downward accountability toward the people who are served. As a possible solution, "Mandatory beneficiary satisfaction surveys are proposed as a means of improving monitoring systems and making programs more responsive to the needs of affected populations."[24]

In a provocative challenge to the relevance and quality of programs provided by the IRC, the former CEO, Reynold Levy, challenged his staff by asking, "If you gave refugees the money, would they buy your services?"

Where Is the Evidence?

Education is a basic human right enshrined in the Universal Declaration of Human Rights, the Refugee Convention and related Protocols, the Covenant on Economic, Social and Cultural Rights, the Convention on the Rights of the Child, the Guiding Principles of Internal Displacement, and the Geneva Conventions.

Alan Farstrup of the International Reading Association stated in a letter that the education of women and girls is at the root of strong societies: "Educated women and girls are empowered to ensure their children get the best education and medical care. They understand the value of good nutrition and lifestyles that provide physical, mental, and emotional health. Education for all citizens is one of the best defenses against the epidemics of sexual exploitation, child trafficking, child labor, and AIDS."

In many of the bleak and hopeless refugee camps around the world, the hazards facing youth are increasingly pernicious. Boredom and the hopeless prospects for the future can hardly compete with the potential excitement of joining their peers in rebel militias.

There are as many as 300,000 child soldiers around the world functioning not only as combatants but as sentries, cooks, sex slaves, and porters. The charm of purposeful and energetic activity in an otherwise stale environment is irresistible. Girls and young women are also lured by the possibility of getting their basic needs met as well as having a sense of belonging that conscription into the multitude of female roles that an insurgency or rebellion can offer.

In a situation of refuge, emergency education programs have demonstrated a measurable decrease in the number of minors conscripted into fighting forces in conflict settings. In November 1998, the Oslo/Hadeland conference on child protection claimed, "Experience shows that education has a preventive effect on recruitment, abduction, and gender based violence, and thereby serves as an important protection tool."[25]

Adolescence is "a time of vulnerability with the uncertainties and turbulence of physical, mental, and emotional development" and accounts for the susceptibility to recruitment in fighting forces.[26] We are forced to accept the difficult truth that many children join armed groups voluntarily rather than through force by abduction or coercion. Together, violence, poverty, and lack of educational opportunity play a pivotal role in seducing youth to join belligerents in a war to which they are in fact victims.

The lure of structure, purpose, and a vague promise of future prosperity are substantial incentives to abandon the dreary life in a refugee camp. When Socrates was given a choice of death or exile, he chose death because he knew that a refugee dies many deaths, and he drank the Hemlock potion that he considered a desirable alternative to the interminable existence of being cast away.

The paucity of options available to refugee youth who have lost their country, family integrity, and future orientation can make illicit activities and other alternatives such as a criminal lifestyle more tempting. Here again, education can be both preventive and remedial. Victor Hugo is quoted as saying, "He who opens a school door, closes a prison."

The philosopher Jean-Jacques Rousseau asserted that if you "educate the women in a society and there will be no need to build prisons." In the Bahai faith, there is a maxim that contends that "if you have money to educate only one child, you educate the girl child." In the United States, the single most predictive factor leading to criminality is low maternal education. There are more African American men in prison at this time than there are in college.[27]

Education is a key determinant of income. Each single year of a girl's primary education correlates with a 10 to 20 percent increase in a woman's wages later in life. This contributes a dramatic illustration of the potential of educational remedies for global poverty. Yet almost two-thirds of children excluded from primary school are girls. Not only do educational opportunities increase the earnings of individual women but education also increases women's participation in their country's labor force. Education is essential for the development of competitive economies and it is also indispensable in the formation of democratic societies. Kofi Annan, former Secretary-General of the United Nations, stated, "There is no tool for development more effective than the education of girls. If we are to succeed in our efforts to build a more healthy, peaceful, and equitable world, the classrooms of the world have to be full of girls as well as boys."[28]

An educated and well-informed population is the raw ingredient for civic engagement, good governance, and the dynamic exchange between society and its leaders in a thriving democracy. This is a crucial component for many of the world's most fragile states. Write economists Paul Collier and David Dollar, "Investment in education has a positive impact on social reform and transformational processes. There is a significant correlation between higher rates of school enrollment and a lower risk of recurrence of civil war."[29]

Can Education Save Lives?

In trying to assert its relevance among the hierarchy of relief interventions, some emergency education practitioners have maintained that although "education might not save lives, education does save minds." This doesn't go far enough.

Education is much more vital than that. Each year of a girl's schooling results in a 5 to 10 percent reduction in infant death. Education results in fewer and healthier children. In "Questioning the Solution," David Werner notes, "Maternal education is clearly associated with children's mortality, in that a child's probability of dying is inversely related to the mother's years of schooling. Maternal education in one of the strongest socioeconomic factors associated with children's survival."[30]

A revealing report issued by the International Save the Children Alliance observes that reaching maturity without basic literacy skills and rudimentary hygiene awareness "can lead to a life of grueling work and an early death." It adds, "Babies born to mothers with no education are twice as likely to die as those born to mothers with three years or more of primary education."[31]

A useful aphorism of global public health observes that there is no such thing as a "tropical disease"; there are only diseases of poverty and ignorance. One of the most important health interventions in a refugee camp does not take place at the desk of a healthcare provider but rather in the waiting room of the clinic. Many health centers and clinics organized by international aid groups exploit the captive audience of the waiting area for "health education animators" to impart lively lessons on healthy behavior like hand-washing, clean water handling, sanitation, and other crucial lifesaving lessons.

Not unlike Western industrialized countries, healthy behavior has a greater impact on morbidity and longevity than any pill, injection, or clinical intervention. One of the most potent remedies that we administer in our clinics in refugee camps is education, hygiene awareness, simple prevention strategies for common illnesses, and the promotion of healthy behavior.

Education in emergency settings is also a useful forum for disseminating vital survival information such as awareness of landmines and unexploded ordnance, tolerance and conflict resolution skills, and even basic rumor control.

Education and Child-Spacing

Many uprooted families, in order to increase the chances of survival, make the difficult decision to have increasingly more children as this becomes their only form of social welfare. The inevitable loss of children as a result of high infant mortality rates forces some families to have even more offspring.

Education offers more alternatives to girls and young women in all societies and exerts a particularly striking role in determining family size. Extending a girls' schooling has been proven to delay marriage and childbearing. In most developed and developing countries, prolonging the opportunities for education decreases population growth. According to a 1995 report by the UN, education has a direct correlative effect on fertility levels whenever it is extended beyond seven years for girls. In other words, by increasing the number of years of maternal education, the number of children in a family decreases.

This correlation is even more dramatically illustrated in the United States. The highest fertility rates are among poorly educated African American girls, while African American women with graduate degrees have the least number of children of any demographic in the United States.[32]

For many women and girls around the world, pregnancy and childbirth are life-threatening events. More than half a million girls and women die each year during pregnancy and childbirth. Ninety-nine percent of these deaths occur in developing countries.[33]

Correlations of family size with years of girls' schooling make a compelling case that education has an indirect but considerable effect on child-spacing. Among refugees and displaced people, this is a matter of life and death. Tribal peoples of the Darfur region of Sudan have an expression that attests to this: "A woman who is pregnant has one foot in the grave." The routine obstetrical complications of pregnancy and childbirth in these impoverished environments can be deadly.

The State of the World's Girls 2007 by Plan International reports, "Worldwide, some 14 million girls and women between the ages of 15 and 19—both married and unmarried—give birth each year. That is 40,000 every day. Pregnancy is a leading cause of death for young women aged 15 to 19 worldwide, with complications of childbirth and unsafe abortion being the major factors." The authors add, "Girls aged 15 to 19 are twice as likely to die in childbirth as those in their twenties. Girls under age 15 are five times as likely to die as those in their twenties."[34]

A recent UNICEF study revealed that a girl in southern Sudan has a greater probability of dying in pregnancy or childbirth than completing primary school: a girl's chances of dying from the complications of pregnancy or childbirth are about 1 in 9 in southern Sudan; her chances of completing primary school are 1 in 100. It notes, "The expansion of female secondary education may be the best single policy for achieving substantial reductions in fertility."[35]

One can only imagine the dangers in a refugee camp where the fertility rates are generally greater than in the regions from where these refugees have fled. It is estimated that as many as 25 percent of girls and women of reproductive age in a refugee camp are pregnant. When you consider that about 15 percent of all human pregnancies— among any society in the world—routinely result in obstetric complications, a simple numerical calculation can reveal the dangers that await girls with poor access to health care and little opportunity for education.

If one were to consider the impact in a settlement of 90,000 displaced people, such as Kalma camp in South Darfur, the lethality of the situation becomes obvious. Opportunities for education can reduce pregnancy rates. Controlling pregnancy rates reduces death due to obstetrical complications. From this perspective, education can be lifesaving.

Might Does Not Make Right

In spite of the overwhelming evidence that education promotes stable, functioning, and prosperous democracies, the total annual contribution for global education from the United States before September 11, 2001 was not even enough to build twenty American high schools.[36]

This is a particularly disturbing statistic when one considers the importance of healthy, educated, and pluralistic societies in the Middle East to the security of the West. Yet the United States has a relatively modest literacy initiative in several Middle Eastern states and makes only token contributions to education and basic schooling in others.

The U.S. government spent more money in one minute of cruise missile attacks in 1998 in Afghanistan than it had contributed over the previous ten years for educational development in that country. There had not been a formal education system in all of Afghanistan in over two decades. It is a cruel irony that the very first U.S. soldier killed in Afghanistan was shot by a fourteen-year old Afghan sniper.[37]

It should become increasingly clear that the victory against terror will not be won on the battlefield. The enemy is not terrorism; the enemy is ignorance and poverty. An effective protocol for peace must include education and development. Terrorism is a symptom of profound cultural and ideological misunderstanding. It is fear of the unknown, contempt for foreign ideas, and paranoia born of ignorance. According to Nelson Mandela, "Education is the most powerful weapon you can use to change the world."

Education and intercultural awareness are critical in changing attitudes, opening people to mutual understanding, and pluralistic ideas. We can never underestimate the danger of ignorance. It is the underdeveloped and unenlightened society that nourishes intolerance and is vulnerable to radicalization and demagoguery. In this war, the pen is far mightier than the sword.

Skills for critical thinking, developed through transformative educational programs, can encourage students to question commonly held assumptions, ethnic exclusion, and malign societal norms. The uninformed and unquestioning mind is susceptible to poisoning by the twisted views of ruthless and persuasive demagogues. As the former prime minister of the United Kingdom Tony Blair has said, "The only way you're going to knock out this terrorism eventually is not just through force of arms but through force of ideas."

It costs approximately $50 to $100 a year to educate a child in a developing country.[38] When one considers the colossal expense of the war on terror, the question is not how can we fund education but rather how can we afford not to? Funding for education is not just a moral issue; it is a question of national interest and global peace and security.

"If we are going to win this war against terrorism, we have to be willing to invest in the lives and livelihoods of the people of the developing world," pronounced Senator Dianne Feinstein following the events of September 11. Yet the United States allocates only 3 percent of its development assistance to educational programming.

Education in Emergencies: A Lifeline

The provision of structured routines and educational programs creates an environment where a child can reestablish a sense of normality, predictability, and purpose. Attendance in a quality educational forum not only provides opportunities for constructive engagement with peers but also creates a reality that is distinct from the traumatic losses and deprivations that led to the displacement.

The UN Inter-Agency Standing Committee warns, "Donor priorities have moved away from funding longer-term development needs to a concentration on disaster relief. This has created a situation where humanitarian assistance is forced to focus on life saving activities and meeting immediate emergency needs."[39]

In order to challenge the shortsighted assumptions that impede the legitimacy of Education in Emergencies among the arsenal of vital relief interventions, we must confront the institutional pressures that drive this phenomenon. In 2004, only about 1.5 percent of the total global humanitarian pledges went to education programs.[40]

In 2001, when only 900,000 students were enrolled in schools throughout Afghanistan, a total of $250 million was spent on education. Now in 2007, with almost 6 million students enrolled in schools in Afghanistan, there is only $80 million available for education.

Of the meager funding for global education initiatives, donors tend to apportion less funding to the countries that need it most. Official development donors allocate proportionately higher funding to middle-income countries (49 percent) than low-income countries (33 percent), and even less (18 percent) to conflict-affected countries.[41]

International financial support for educational programming should be available at the onset of a crisis. Learning environments should be secured early in a crisis that can support the physical and emotional well-being of children and adolescents uprooted by conflict or disasters. Integration and recognition of the validity of both refugee students and refugee teachers must be promoted in host countries. The INEE's *Minimum Standards for Education in Emergencies, Chronic Crises, and Early Reconstruction* should be the international benchmark for measuring the success of this crucial intervention for children and adolescents affected by war and disasters.

Conventional short-term funding cycles as well as lack of objective assessments of what refugees need continue to promote knee-jerk responses to crises. To alleviate these shortsighted traditions, we must work with major donors to adopt multiyear funding cycles and avoid the crisis-driven ethos that currently permeates the humanitarian culture.

Only when we examine the assumptions, attitudes, and policies of international assistance will we be able to advance a culture of critical reflection and sincere analysis within the aid community. Education in Emergencies is a lifeline for children and youth imperiled in a bleak and interminable exile. Ultimately, we must be forced to ask ourselves: are we prolonging life or are we just postponing death?

9 | The Power of the Curriculum

FALK PINGEL

Teachers need tools to transmit knowledge to their pupils and to furnish them with competencies. As a rule, in emergency and post-conflict situations the pedagogical equipment is poor and the material conditions appalling, with school buildings destroyed and classrooms looted. A classroom may often have to accommodate forty or more youngsters of quite different learning abilities. Pupils need material they can rely on and make use of for homework and repetition. Textbooks and additional printed materials are still the most important and often used tools, representing contents and methodology of what should be learned—in spite of the increasing role of newspapers, documentaries, and the Internet. Although new media have complemented the traditional set of learning tools in the past decades, they have thus far not replaced them and, additionally, they play a minor role or are not even available during emergencies.

In the broader meaning, all these tools, including a teacher's professional capacity to attract the pupils' attention, represent the classroom or school curriculum. However, schools are not isolated monads, functioning peacefully while the sociopolitical surrounding is shattered. In well-off countries with stable institutions and a highly developed communication infrastructure, "school autonomy" has become a catchword, meaning that schools should be free to develop their own curriculum. Conversely, in conflict-ridden areas where communication lines are often interrupted and institutional structures are weakened, schools are bound more strictly than ever to observe policy guidelines, and they have less room for maneuver to construct materials on their own according to local needs and available resources.

Education in general and the teaching of history and the social sciences in particular, are not "neutral." The content that is taught and the moral and political messages that may be associated with it are often related to the conflict. Through the interpretations delivered—implicitly or explicitly, unwittingly or on purpose—teachers take sides in the struggle and can be amenable to criticism from pupils, parents, and authorities. As they are exposed to endangered situations, teachers need legitimacy for inculcating facts and views in pupils. The observance of the curriculum gives them a feeling of security and "political correctness" in times of insecurity, turmoil, and change.

The term "curriculum" has different dimensions:

- The overall "philosophy of education" refers to the values and general objectives of education.
- Against this broad background, the relevant authorities create an idea of the "intended curriculum."
- This serves as a guideline for the formulation of the "formal" or "official curriculum," the approved plan for instruction containing the various syllabuses that represent the concrete content that teachers are obliged to teach.[1]
- To what extent the formal curriculum differs from the intended one depends on the political procedures that regulate the work of the curriculum developers and the mechanism of approval. In a democratic society the formal curriculum may considerably differ from the intentions of the government as curriculum committees are made up of a number of experts and stakeholders' representatives who try to execute their own agenda. In authoritarian or dictatorial states, the curriculum strictly reflects the political intentions of the ruling group.
- Nevertheless, in authoritarian regimes, the "hidden curriculum," what teachers actually teach, may be different to a certain extent from what is described in the formal curriculum wanted by the government.

The curriculum is used foremost as a pedagogical instrument, providing guidance for selecting the content and methodology of the teaching process. However, it is also an instrument of power, by which, through the formal curriculum, the state defines what should be taught and how. Particularly in post-conflict situations, when a new pedagogical philosophy is to be formulated, this power aspect is of crucial importance. Who participates in the construction of the curriculum? Does the curriculum reflect the cultural and social makeup of the society?

This chapter will examine the restrictions imposed on teachers and pupils through the curriculum, as well as the opportunity it offers for coping actively with conflict situations. Examples are drawn in particular from interventions and consultancy activities in the field of education performed in the course of projects conducted by the Georg Eckert Institute for International Textbook Research. Although it takes a comparative approach, the chapter focuses on conflict and post-conflict areas and issues such as the former Yugoslavia, Palestine, and South Africa, where I gained my personal experience in the topic. Furthermore, the chapter concentrates on the challenges that the formation of a post-conflict collective identity, including its historical dimension, poses to teachers, textbook writers, and curriculum experts. Therefore, those subjects that mainly deal with these topics— the humanities and social studies in particular—will be taken into consideration.

War and Its Impact on Post-war Curricula

Studies have shown that the particular conflict itself needs careful analysis when defining the strategies to deal with its challenges and after effects in the classroom. The more violent it is has been, the stronger its consequences for the revision of the education system. In a war, the very existence of a society is at stake. The more threatening the enemy is, the more forces are mobilized for the fighting. The military front is supported by an ideological one. History must prove that the war has been "just," based on claims and rights that are anchored in the past and dignified with the name of tradition. The demands that led to war are, of course, mutually exclusive; they cannot legitimately be shared by the "other" side. The wartime curricula of all sides are imprinted by strategies that prove the legitimacy of one's own actions and refute the illegitimate claims of the adversary.

The dichotomies of right and wrong that war produces do not disappear with the end of the conflict. On the victorious side, one finds myths of superiority and righteousness. The values that supported warfare take on an almost absolute significance after the event itself and will be commemorated in the future. On the side of the defeated, feelings of "victimization," unfair suppression, and collective guilt may arise and generate strategies to exculpate themselves. The images of self and other on both sides of the conflict do not match. In the precarious situation after civil war, victors and vanquished must seek ways to live together and to reconcile worldviews that up to now excluded and fought each other. Education is charged with the difficult task of delivering to the young generation new views of the world that explain the conflict but do not perpetuate it. By definition, post-conflict educational philosophy is related to the conflict, whether it strives to ignore, neutralize, or openly address it.

The notion of righteousness and the missionary consciousness to civilize the post-war world are seen in the U.S. and British textbooks published after 1945.[2] As such interpretative patterns harmed a wanted ideological rapprochement with former enemies, joint textbook consultations conducted under the umbrella of UNESCO, governments, or NGOs after the Second World War had an impact on the curricula and textbooks of the victorious side as well who had to "normalize" its exclusive claims on representing the historical truth.

The German-French textbook talks in the 1950s were a significant example of cooperation on equal terms between scholars, teachers, and textbook authors of both sides. At times, the myth of Versailles, that Germany was the only culprit for the outbreak of the First World War, impeded a common interpretation of German-French relations in the twentieth century (as there was no doubt about Germany's aggressive role in 1939). A careful reinterpretation of pre– and post–First World War history stood at the early development of joint textbook consultations, which continued up to our times and resulted in the first binational, curricular history textbook ever published (Le Quintrec/Geiss/Bernlochner).[3] The German-French textbook recommendations set the First World War in the wider context of imperialism and the struggle for domination between the great powers in which France also had its share. It took, however, two further decades to give collaboration its appropriate place in French history textbooks alongside the hymn of resistance that, according to the schoolbooks of the post-war area, had been the only reaction of the French population to Nazi occupation.

The U.S.-Canada consultations are an example of textbook talks between victorious powers meant to overcome traditional notions of superiority on the one side and of inferiority on the other. Their joint recommendations in 1948 stressed the mutual political as well as economic dependency and cultural interrelatedness between the two countries and acknowledged the contribution of Canada to the victory over Nazi Germany combating the prejudice of Canada being just an appendix of the mightier United States.

War creates not only its own myths, but also new intranational and international power balances. Both have to be reflected in the curricula in order to deepen the understanding of conflicts and develop awareness for peaceful conflict management. Bilateral or multilateral textbook consultations between the former combatants proved to be an appropriate tool for reaching this aim after the Second World War. However, they were mostly conducted between former adversaries that afterward established peaceful relations and even partly became political allies. A similar conducive political framework is missing in the majority of recent cases of protracted internal conflicts, as in Africa and Latin America or ongoing warlike interstate violence, such as between Israel and Palestine.

Long-lasting repercussions of war molded the educational objectives of all states that emerged from the former socialist Yugoslavia. Bosnia and Herzegovina might be one of the most salient examples of where the war put a durable mark on the curricula. The teaching programs emanating from the local authorities during and immediately after the war of 1992–1995 became instrumental in perpetuating the conflict even after actual fighting was over. Education has served as a crucial tool in defending the cultural borders that the war had created and transferring the conflict from the political and military arena to the cultural sphere.

The war of the 1990s brought the common educational system inherited from the Yugoslav area to a breakdown. During the war in Bosnia-Herzegovina, each of the three different political units—the Serb Republic, the Croat-dominated Herzegovina, and the rest of the newly declared independent Republic of Bosnia-Herzegovina, with its heartland around Sarajevo—introduced its own curricula and textbooks that stressed differences rather than commonalties between the three communities, accusing the "other" of suppressing one's own culture and waging aggressive military actions amounting sometimes—not only at present, but also in history—to "genocide."[4]

As the Croat and Serb areas imported their schoolbooks from the neighboring states of Croatia and Serbia, which had been involved in a bloody war as well, these books were designed to foster Serb or Croat identity respectively. They neglected to discuss the idea of an independent Bosnia-Herzegovina. Confronted with this situation, the government in Sarajevo, with a Bosnian majority, developed their own textbooks and syllabuses that for the first time propagated an independent, democratic state of Bosnia and Herzegovina. As the Dayton Peace Agreement entrusted the political units of the united, though highly federalized, if not compartmentalized Bosnia and Herzegovina—the Serb Republic and the Federation of Bosnia-Herzegovina with its subunits, the Cantons—with the responsibility for education, the separated educational system survived the war and could not function as a tool for developing an overall national consciousness that might help overcome the ethnic hatred that had permeated the ideological outlooks of the country's population during the war. Attempts from within at harmonizing the different curricula and textbooks and screening them for any apparent disparaging, discriminating passages and biased presentations failed because the leading politicians (mis)-used their autonomy in handling the curriculum only for the alleged sake of their own community.

Only five years after the war, the international community monitoring the implementation of the peace accords intervened and initiated a program of curriculum and textbook revision because racial, ethnic, and religious discrimination was regarded as an offence against the Dayton principles. Even now, fifteen years after Dayton, the revision process is still going on. Appropriate local legislation that should enforce antidiscriminatory education principles on all levels in Bosnia and Herzegovina had partly to be imposed on the local authorities by the international community.

All parties involved in the struggle for or against an education stressing ethnicity and cultural difference used the power of the curriculum to reach their aims. For the ministries of education it was the main tool to uphold an ethnically separated school system. But the international community and reform-oriented local politicians also used the curriculum as a lever for chance. The first leading reform document, *A Message to the People of Bosnia and Herzegovina*, envisaged a harmonization and revision of all curricula according to principles of nondiscrimination and multiculturalism. Although the legal process for changing the formal curriculum proved to be long and not without setbacks, practical steps forward were supported by teachers who followed a hidden curriculum, believing in a indiscriminately mixed education as they knew it from Yugoslav times.

In addition, somewhat disappointed at the slow pace of the country's political institutions, the international community put more emphasis on teacher training to develop capacities for bottom-up reform. This approach was furthered by a number of local expert teachers, textbook authors, and reviewers who worked in ethnically mixed curriculum and textbook commissions. Many of them became the most ardent and persuasive advocates of reform. It is hoped that in a long-term perspective a reform-friendly formal curriculum will meet with a reform-driven hidden curriculum and that both together will create synergies that will alter the whole system in a sustainable way.

In all socialist/communist countries in central and Eastern Europe a strong curricular orientation prevailed. The transformation of the political system did not cast into doubt the dominance of the formal curriculum, which was to become the main pillar of change. Three major curriculum reform steps can be discerned:

- The first step toward educational change began when state frontiers were opened up. Until then, textbook authors and curriculum planners had had little opportunity to communicate with their colleagues in the West. Although the ensuing personal contacts were important in promoting an atmosphere of intellectual curiosity and a readiness to adopt new teaching methods, they had little direct effect on the educational systems as a whole. In general, this first phase of reform was marked by negative or defensive rather than positive and constructive measures. Much of the material that was clearly formulated on Marxist principles was removed, but the overall content and the actual subjects taught were not called into question. There was also little change in the teaching methods employed. The canon of subjects was hardly altered at this stage with the exception of value-loaded disciplines that were either forbidden (such as instruction in Marxist theory) or reintroduced (such as religious instruction).

- The second reform phase led to an overall revision of the curricula and the introduction of teaching methods that were more discursive in style than the rote learning and memorization practices of the past. Like its predecessor, this reform was normally implemented by a centralized ministry, which clung to the system of a uniform official curriculum and prescribed the same textbook for use in all schools in the country. Although so-called alternative material became available, it did not replace the one and only officially approved textbook.

- Only the third stage of reform brought a fundamental change in teaching methods and textbook content. The subjects taught were no longer viewed in isolation but grouped together on an interdisciplinary basis so as to allow pupils a certain amount of choice in what they learned. At the same time, both teachers and pupils were given greater freedom in their approach toward a particular subject and schools got more leeway in implementing the official curriculum. A competitive textbook market flourished ending the state monopoly on the production of schoolbooks.

These phases of the general institutional reform can be matched with the following changes in the philosophy of education:

- The Marxist interpretation was superseded by a more backward-orientated, teleological interpretation of how the nation-state was formed. Instead of the socialist international dimension the individualism of nations was now stressed. States rediscovered themselves as national entities and rewrote their history to comply with the idea of an ethnically homogenous nation-state.
- The orientation toward nation-building or renationalization was amended by a growing European awareness. Europe served as a point of reference for a more inclusive identity concept, although in the social sciences the national and political dimension—as compared to culture and economy—prevailed.
- A cultural and social element (with an emphasis on everyday experience to supplement or replace the traditional focus on an institutional approach) began to appear alongside the political dimension, which was increasingly underpinned by European and international/global aspects. Textbooks started to focus on the internal groupings and subdivisions of the particular society, a development that was due, to a considerable degree, to the pressures and recommendations emanating from European organizations. In many countries this shift was accompanied by a public debate, at times heated, on the aims and objectives of teaching. There was still a strong feeling that pupils should learn to identify with the state in order to counterbalance the multicultural approach and the emphasis on the European and international dimension.

Some countries, such as the Baltic states, Poland, the Czech Republic, Hungary, Slovenia and Bulgaria, experienced these phases as a relatively smooth transition driven not only by state institutions but also by new professional organizations and NGOs. In other countries, where the political transition was accompanied by internal conflicts, regressive developments took place—often initiated by governmental interference (as in Romania, Slovakia, Serbia, and Croatia). Progress was hampered, above all, where the road to national sovereignty was blocked by persistent armed conflicts or where ideological disputes concerning the kind of system that should replace communism led to repeated changes in the overall educational objectives.

However, by implementing EU norms and institutional structures common in the EU, such delays and setbacks could be successfully overcome. The integration into a wider international framework offered security and expert guidance at the same time. Organizations such as the Council of Europe and the EU could also offer criteria and objectives for reform, which were already implemented in their member countries.[5]

Hypothetically, some general conclusions can be derived from the more or less clearly sequenced stages of reform in the former socialist world:

- The apparent will of change was accompanied by a strong desire to establish a new focus of political identity. This gave curricula and textbooks a new slant, but it did not much change the structure of the content, which was just interpreted in a different way. Interestingly, the models used for a first revision in this phase were textbooks and curricula from the 1920s, when most of these states enjoyed for the first time full sovereignty. Education for national unity was the leading paradigm that replaced communist internationalism.
- On the basis of this newly secured identity concept new methods of teaching were adopted whereby contacts with Western European colleagues already established in the first phase played an important role. They helped to start a controlled trial phase that opened up the narrow national perspective.
- Realistic hopes for the future helped to widen the horizon and even face the challenges of a common European educational sphere, as well as globalization. This advanced the innovation of methodologies, and led to a differentiation of content issues with the ultimate aim to develop an overall curriculum that accepts differences and acknowledges the increasing multicultural makeup of modern European societies. Of course, this process has not yet ended and is not unchallenged, even in Western Europe, but the paradigm of a homogeneous nation is no longer the exclusive focus of identity politics in education.

• Although the reform process was steered by the governments, professional associations and NGOs had a say as well; they delivered ideas and concepts and were represented in the curriculum committees. Only in the third stage did the curriculum become less prescriptive and defined broad content areas and a variety of methodologies that gave teachers a choice to tailor their lessons according to the interests and learning abilities of their pupils and offered schools the opportunity to develop their own profile.

The detrimental effects of war on a smooth transition process became obvious in Serbia, Croatia, and Bosnia-Herzegovina, where wars impeded or delayed the second stage of reform. The supply of books and equipment deteriorated significantly as a result of the destruction and lack of investment during and after the war. The methods and contents inherited from the socialist period remained in place much longer, with the only difference being that the nationalist-ethnic dimension was given much greater emphasis.

Among the war-shattered countries, it was Croatia that made the most progress in the textbook sector after the war. At first, the curriculum primarily stressed the continuity of a Croatian-Catholic national history from the Middle Ages onward. The country experienced a historical and political debate in which the nationalists and conservatives wanted to force schools to follow this interpretation. Under the Tudjman government, the nationalistic forces became so influential that some textbooks even legitimized the fascist Ustasha regime as the first modern Croatian state. Nevertheless, multiperspectival, even critical interpretations, which deviated from the prevailing idea of history, have been formulated and have also been integrated into textbooks. This is the result of work carried out by internationally recognized historians at the University of Zagreb who, with their colleagues abroad, participated in the revision of schoolbooks.

In Bosnia and Herzegovina, Serbia as well as in Macedonia and Kosovo, the violent and continuing conflicts led to a hardening of ethnic and national positions and the insistence on one-sided interpretations. Although in Serbia, in the post-Milosevic era, the issues were debated more openly and the Ministry of Education announced that it would promote the development of new textbooks and curricula, the moves toward reform are still hampered.[6]

The transformation process in Eastern Europe can be seen as a backdrop against which the challenges and dilemmas of transition in areas of less favorable politico-economic ramifications can be better understood. In most of the conflict-ridden societies and fragile states in Africa, and partly also in Latin America and Southeast Asia, the conditions that allow the sequencing of the reform process are not given.

- As many societies here look back at a civil war marked by strong and opposed ideological beliefs the post-conflict society cannot yet draw on an accepted model for national cohesion as most of the former socialist states did; rather, education is charged with the task to invent and to instill a new sense of belonging and an awareness of joint responsibility.
- Governments resulting from a civil war have often not yet stable constituencies and tend to bury rather than lay open still existing differences in opinion and traditions; they propagate an ideology of harmony before the contentious issues can be clarified and a consensus can be reached.
- From the very beginning, education is overloaded with the task to heal society without being furnished with the appropriate medicine to do so. The state tends to either prescribe and impose a new curriculum or declare a moratorium on the teaching of sensitive subjects until society seems ripe for a widely accepted consensus. In the first case, teachers are hesitant to wholeheartedly accept and implement the curriculum; in the second one, they feel lost and lose trust in the government and believe that their authority in relation to their pupils will be diminished.
- A stabilizing regional framework similar to the European organizations is missing in almost all African or Latin American countries suffering from civil war or social unrest. Nevertheless, help from outside is needed, although international assistance can often take on the meaning of external intervention.

- Under the pressure of scarce resources and unbearable living conditions of parts of the population, donors and expert organizations often have to decide whether to invest primarily in material reconstruction and institution building or in democratization and liberalization.

Institutionalization Versus a Civil-Society Approach

From a technical point of view, rebuilding institutions and premises like schools seems to be more basic than, for example, teacher training.[7] If, however, prioritizing the material and institutional dimensions of reconstruction means, in practice, reinstitutionalizing old contents and traditional behavioral patterns, the renewal of the education system as a whole cannot succeed. This was exactly what happened in Bosnia and Herzegovina, where the international community concentrated on material reconstruction and institution building in education in the first years after the war, whereas the sole responsibility for defining the content of education lay with local institutions. The institutional approach supports, as a rule, the reestablishment of former elites and/or favors the ruling group; it tends to result in a narrow, prescriptive educational program that hardly pays tribute to a variety of cultural traditions and political opinions.

Curricula-orientated reforms often prolong the process of renewal because the formal procedures for setting up curriculum committees take time, as does the work of the committees. When the Palestinians took over responsibility for education from the Israeli, Jordanian, and Egyptian authorities after the Oslo agreement, the Palestinian National Authority stuck firmly to a clear sequencing of the reform process:

- Construction of new curricula
- Writing of new textbooks that correspond to the new curriculum.

Each year, syllabuses and textbooks for two grades would be implemented with the consequence that Jordanian and Egyptian schoolbooks had to be used in some grades for many years to come and could not reflect the great changes Palestinian society underwent since Oslo. As the textbooks were oriented strictly on policy guidelines, they dealt with contentious issues of the Israeli-Palestinian conflict, like borders and the Oslo peace agreement itself, in a very cursory way. Although this approach was founded in a strong curriculum-based teaching tradition in the Middle East in general, it also allowed the government to exert control over the whole reform process at any step and to limit the influence of grassroots project work.[8]

However, the alternative approach does not produce viable results, either. In areas of protracted, violent conflict NGOs and international institutions put efforts into maintaining basic education for all in order to train the younger generation for better times to come and keep them away from the pressure or temptation of war gangs. This work is often done on a project basis with limited financial means. As a rule, the organizations engaged in this work follow various curricular approaches that are often not harmonized and are implemented only in a particular region of the whole country. An institutionalized durable curriculum is going to be implemented only when the violent conflict is over and stable institutions have been built. Whether the short-term, project-based work at the beginning is able to later imprint itself upon the official curriculum is an open question. It depends largely on:

- How many teachers, textbook authors, and curriculum specialists can be reached,
- How intense the training is, and
- Whether the trained experts will participate later on in the development of the official curriculum.

However, conditions that favor a long-term impact are often not available. The empowerment of educators to teach without curriculum and narrow guidelines, and to use their room of maneuver, is limited by poor teachers' education and low salaries that often force them to have a second job, with consequent lower investment in their own professional advancement. Furthermore, if the curriculum process starts only in the second phase of reconstruction—which is normally the case—it coincides with the handing over of responsibility to the national authorities, so that international institutions and NGOs have limited influence on the make-up of the curriculum committees. However, if the group of empowered experts is strong enough, they may work with professional organizations, parents' representatives, and the like to develop alternative approaches and weaken the impact of the official curriculum by employing a hidden curriculum.

Brian M. Puca has shown, in an impressive study about the American reeducation policy in Germany after the Second World War, how trainings that US institutions offered during the occupation changed attitudes and ideological outlooks of the participants in a sustainable way. Their orientation toward democratic values and active participation in society even survived the period of the 1950s, when conservative trends dominating in the young Federal Republic reversed many innovations in curriculum construction introduced with the help of the US occupation power. This kind of groundwork established an informal curriculum that worked to a certain extent against the official educational philosophy and diminished the impact of the formal curriculum.

The case of Japan shows some salient differences to that of Germany. While at the start of the occupation the United States applied almost the same strategy to the two countries, it changed its policy almost totally in the wake of the Cold War and the Korean War. The Japanese government again gained full control over curriculum development and textbook approval, which, in contrast to Germany, set a narrow frame for teachers and authors and was able to suppress for a long time any critical questioning of Japan's role in the Second World War.[9] Though other voices were never silenced, they were marginalized and restricted to left-wing opposition groups. In particular, teachers unions have remained much more open to a reconciliatory approach than the government, but the narrow curricular frame and the orientation toward rote and exam-driven learning have restricted their room to maneuver in the classroom considerably.[10]

During war, centralized administrative structures often break down. Where the education system is heavily dependent on such structures, regular schooling comes to an end. To continue, individual initiatives and the will to take over responsibilities on a local level are needed. In and around Sarajevo, the government of Bosnia-Herzego-vina encouraged teachers to give lessons at home or in makeshift ac-commodations when classrooms were destroyed and to be inventive in producing their own learning material or make use of any media at hand where schoolbooks were no longer available. Under the pressure of the Serb siege, an informal war curriculum emerged that fostered the ideas of self-defense and victimization. As mentioned above, it also influenced new official textbooks that were developed during the siege and mostly used well into the post-war decade. So, to a certain extent, the informal wartime curriculum turned into the formal post-war teaching plan.

Teaching in out-of-school-environments, under the constant threat of being bombed or looted, leaves a deep impression on teachers and pupils that is not forgotten when the actual fighting is over and school life returns to normal. The feelings of that time, particularly hatred against the aggressor, likely find their way into the post-conflict infor-mal or hidden curriculum, but often also form part of the official curriculum.

In conflict-shattered regions where legitimate political institutions no longer exist, the informal wartime curriculum may be influenced in the worst case by local warlords, in the best case by international peace organizations. The former preconflict curriculum may be used in parallel to the emerging new informal interpretations and method-ology or, more likely, permeated by them step by step. This mixture of traditional elements, war ideology, and an orientation toward peace and mutual understanding normally represents the point of departure for the reinstitutionalization of an official curriculum in the post-con-flict period.

To work hand in hand on material reconstruction, institution building, curriculum reform geared at democratization and human rights education, and on the empowerment of professionals is almost impossible. It is due to the short-term perspectives of donors that external agents of reform are eager to produce quick results. The World Bank report of 2005 frankly states: "In the heat of international pressure and community demand for textbook and curriculum devel-opment" actions "get out of sequence" and "exceed the capacities of local authorities to implement them."[11] What can be done, despite this almost inevitable dilemma to prioritize, is to sharpen the awareness of the shortcomings of any approach in order to develop a coordinated education policy as soon as possible and reduce the negative impact of limited choices at the beginnings.

The Curriculum in Transition: Negative Intervention Versus Constructive Renovation

In the period following a peace agreement or ceasefire, the political forces in power have a wide choice of strategies to feed their ideological and value orientation into the education system. As schools should start to operate again as soon as possible, so-called emergency measures have to be introduced in order that the old political ideology not be inculcated in pupils' minds. Since new curricula and textbooks cannot be developed all of a sudden, the first phase of reconstruction in education is often marked by "negative" intervention. The most radical measure is to stop the teaching of subjects that are more or less infected as a whole by the old ideology. In addition, teaching material that eulogizes the old system is withdrawn or screened for objectionable formulations, illustrations, and the like. Imported books and ready-made teaching plans from outside may substitute the withdrawn ones. This may work for the sciences, but it is highly problematic in the humanities and social studies.

It is the precarious closeness of subjects like history and civics, or "moral" education, to the conflict at stake that often leads politicians in emergency and post-conflict situation to the decision to withdraw them from the curriculum completely. As long as no new interpretation of the contentious past is available and no viable power-sharing agreement has been reached, the authorities tend to either mistrust the teaching of such subjects—since these disciplines seem unable to contribute to the stabilization of the shattered society—or use them for legitimizing their own position and diminishing the role of the adversary. In the past, the occupation of Germany and Japan served as an exemplary case of how to cope with the transformation of a militaristic, authoritarian, or dictatorial system into a pluralistic democracy. In both countries the occupying powers resorted to this kind of emergency intervention immediately after having taken over control; the teaching of history was suspended for almost two years.

Another emergency measure was recently applied by the U.S. military administration, with the support of UNESCO, in Iraq where all textbooks were checked and obvious references to Saddam's dictatorship and its ideology removed. This approach makes educational material available quickly, and it allows teachers to work with devices that they are familiar with. It is, however, interpreted by large parts of the population as an imposed intervention of the new rulers (whether foreign powers or a new internal ruling group) without consultation with teachers' or parents' organizations. It may be tolerated in the immediate reconstruction phase, but if it continues when the social situation has settled to some extent it likely provokes open or silent resistance from teachers and parents. This happened in Bosnia and Herzegovina when the international community became aware that the focus on material reconstruction and reinstitutionalization of the education system on a local basis had led to firmly established discriminatory syllabi and textbooks. As a first step of fast intervention the international community, under the leading of role of UNESCO, blacked out so-called objectionable material in textbooks already printed so that it was obvious for all users that the international community had forbidden content to be taught that was formerly approved by the local educational authorities.

In the immediate post-conflict phase, the authorities are caught between the Scylla of quick and radical intervention and the Charybdis of protracting the educational objectives and methods of the old system, which should be overcome. It is contested which strategy leads to better results. The World Bank report expressly speaks out for acknowledging "the importance of symbolism in education . . . that signal that the reform has started."[12] Rose and Greeley join them admitting that in the beginning one-sided measures have to be taken, whereas later on a more compromise approach should be chosen.[13]

As useful as it may be to make immediate changes visible, misunderstandings can arise. The World Bank Report refers to "textbook purging" as one of these recommended symbolic acts. However, as we have seen, imposed textbook screening can provoke resistance rather than stimulate active participation in the reform process.[14]

In addition, the minimum standards for education as developed by the Inter-Agency for Education in Emergencies (INEE) make this problematic dilemma obvious, contending that curriculum development "can be a long and difficult process" in which stakeholders should be involved and "a range of actors . . . including learners, community members, teachers, facilitators, educational authorities and programme managers . . . consulted." In fact, because of pressing needs for action and results, "curricula are often adapted from either the host country, the country of origin or other emergency settings" that offer only scant chances for wider participation of local interest groups and experts.[15] INEE warns that innovations should be introduced slowly; with respect to subjects that may address the conflict directly, they recommend for the first year of intervention to "initiate discussion on history and civic education, knowing that curriculum building will take time in these sensitive but important subjects," then "assist curriculum developers to go beyond syllabus/topic listings . . . to trialling lessons and to developing teachers' guides and theme/topic materials for pupils." This means to make teachers more autonomous and less dependent on the written curriculum. Nevertheless, it is important to invest in the development of written/printed material as soon as possible so that teachers have something on hand that they can show to others, but that is open to revision.

The INEE guidelines state, "in the light of experience . . . without teaching materials, teachers will simply not teach difficult or sensitive topics." Instead of a problematic revision of old textbooks that are no longer appropriate for the post-conflict situation and have lost legitimacy, trial lessons and experimental teaching material can bridge the gap, as long as more suitable teaching material is not at hand. This is also corroborated by UNESCO's guidelines on emergency education, although not implemented in all projects supported by it.

Conflict-torn societies with a federalized education system have a better chance to cope with quick provisional changes because they can be reestablished more easily and more quickly than centralized structures and can respond in a more flexible way to differing local educational contexts. NGOs, particularly smaller ones, act often only at local level and draw on civil society groups or engaged individuals who are ready for change. They may also act if governmental support or approval cannot be reached. Larger international institutions such as UNESCO or the World Bank are bound to conclude agreements with the authorities before they can implement a program. The bottom-up approach adjusts faster to local needs and can combine the establishment of new teaching devices and contents with teacher training on the spot; it comes in crisis, however, in the next phase of reform when countrywide and official structures are being reestablished.

Teachers are hesitant or even afraid to introduce material and ideas that are not officially acknowledged.[16] This is particularly true for teachers who are used to working in centralized structures or under authoritarian rule. Waiting for official guidance, which often is not available for a period of months or even years, slows down the reform steps that NGOs, engaged individuals, and civil-society groups want to introduce. Therefore, the strengthening of teachers' authority and the development of their professional pride, as well as their standing with the government, should be one of the aims of teacher training. However, in some situations teachers can fall under serious pressure from local rulers who want to devise their own policy challenging the officially acknowledged authorities.

In a training seminar conducted at the Georg Eckert Institute, teachers and headmasters from Iraq felt that the central government in Baghdad provided the only trusted and objective guidance in curriculum matters, since otherwise they would be at the mercy of unreliable local rulers. However, because communication lines were often broken and the central government was hardly able to provide guidance because of the endangered security situation and internal power play, the reform stagnated. Currently, projects are being developed outside of the country that offer reform steps designed according to the latest curriculum theory, but not yet adapted to the country's unstable and fragmented situation.

Teachers need allies. Bottom-up initiatives taken by engaged professionals, local and international NGOs, and courageous teachers competed with governmental policy in South Africa when it came to the abolishment of the apartheid curriculum in the first half of the 1990s. They started to conduct workshops on how to conceive a new, all-embracing South African history. This, in turn, triggered a broader debate on the multiethnic and multicultural mix of South African society and posed the question of how the various traditions of different communities could be integrated in one curriculum.

At first the government showed interest in this movement, without becoming involved in its activities. However, the longer the debate lasted and the more alternative material appeared, the government changed its attitude and expressed concerns that the process could bring about a variety of different approaches and interpretations instead of fostering a common understanding of South Africa's difficult heritage. This concern appeared to be all the more relevant as responsibility for education is divided between the central government and the provinces, whose authorities partly strove to stress their own cultural traditions and historical roots so that the focus on commonalties seemed in danger of fading away.

In order to avoid painful debates about the contested past, the central government designed a countermeasure. It presented a hastily constructed new curriculum that downgraded history as a separate subject and integrated it into an interdisciplinary social-studies concept that favored social competencies. Most of the nongovernmental initiatives came to a halt. The new formal curriculum triumphed over attempts to develop an informal curriculum that was partly tainted by the flavor of revolutionary anti-apartheid fighting and partly marked by enlightened liberal human-rights positions and so reflected the controversial internal political and academic debate. This informal curriculum invited pupils to discover the past and think about a common future independently.

But such openness was not wanted at a time when the government was trying to concentrate all its efforts on the modernization of society and the economy and therefore favored a curriculum that prioritized sciences, information technology, and social techniques, not critical thinking and problem-orientated discussion. It did not come as a surprise that the new outcome-based curriculum putting competencies and not content to the fore was not well understood by teachers and not accepted by textbook authors, and its implementation was not supported with sufficient resources. New textbooks reached classrooms slowly and apartheid-era history textbooks could still be found in many schools until well into the new millennium.[17]

Governments that fear that a historical debate about sensitive issues related to problems of national pride and self-image may weaken the reconstruction process willingly embrace outcome-based curricular concepts that require pupils to develop competencies and that define contents only in broad terms. However, competencies can be trained only through dealing with concrete issues. If topics related to the conflict are not mentioned or detailed in the curriculum, it is unlikely that they will enjoy a prominent place in teaching. Furthermore, when resources for appropriate in-service training are not provided, modern teaching devices not available and classrooms overcrowded, these modern curricular concepts are doomed to fail. In fact, they either deprive lessons from topics young people are most interested in, or they leave it to the teacher to solve the problem of if and how controversial issues should be dealt with in the classroom.

When Kosovo had thrown off the pressure of Serbian domination but had not yet reached full independence, UNDP provided the country with a modern curriculum that was meant to foster democracy education and to develop the readiness of young people to participate actively in society. It was deliberately shown to international experts to prove Kosovo's readiness to join the international community as a responsible and free state.

The syllabuses, however, neglected the propaganda of narrow ethnic and nationalistic values that were widespread at that time in the media; it did not address the total separation of the Albanian and Serbian school system in Kosovo. To prepare the country for independence, UNMIK handed over responsibility for education to the Kosovar authorities, who used this new opportunity to adapt the history lessons to mainstream Albanian nationalism. Textbooks from Albania needed to be only slightly changed and amended to serve this aim. When UNMIK checked the books produced by the Kosovar ministry of education, its officers were embarrassed by the nationalistic, sometimes mythical content that replaced the former Serbian narrative with a hymn to the brave Albanian national character and its achievements in history.

In this case, the informal curriculum represented by the media, public debate, and the new textbooks was much more influential than the official formal curriculum designed with the assistance of an international organization. Only after the most crucial political questions had been settled and independence no longer questioned did a slow process of textbook revision come into motion. Other international organizations such as the Council of Europe and OSCE are now engaged in teacher training to implement teaching methods as required by, but not yet implemented through, the official curriculum.

Taking Kosovo as an example, Marc Sommers and Peter Buckland write of "parallel worlds" that are created by different organizations implementing distinct objectives with differing methods.[18] This mix of governmental policy, local initiatives and international assistance is a normal feature of transformation processes. Although it might not lead to an overall systematic approach, the mixture may be more appropriate in the end than a unified, streamlined policy to balance out competing concepts of education that are characteristic of countries in transition. In her penetrating analysis of coping with a post–civil war situation in Guatemala, Elizabeth Oglesby showed that the determined curriculum and textbook policy of the government partly failed because implementation was impeded by one or the other political faction of the country.[19] Fresh input had to come from grassroots activities to further advance serious discussion of the conflict in classrooms.

Consensual Curriculum Policy

As these examples have shown, the pace of education reform, the profoundness of change, and the acceptance of new approaches depend to a considerable extent on the role the curriculum plays in the set of reform measures, and in particular on the impact it has on the teachers. Besides the technicalities of selecting contents and methods, the overall aim of curriculum development in post-conflict situations is to invest the fragile society with basic social and political beliefs that allow it to set its trust and hope in state institutions, and to contribute to social cohesion that is mostly lacking after internal conflicts. It is not only that the contents and methods that should be taught receive legitimacy through the official curriculum; the curriculum also gives legitimacy to the government. It justifies its actions and calls upon the pupils and teachers for solidarity and identification with the state.

To find a new, viable and agreed upon concept of history, civic education or a literature canon that reflects the many cultural traits of a society is an honest aim. But it must also serve the purpose of strengthening the awareness of interrelatedness and mutual dependency within a multifaceted society and develop the willingness to cooperate. It can also, alternatively, be used to exert cultural dominance. A curriculum-driven policy can lead to quite different results. If the government has a strong, uncontested position and teachers feel obliged to follow the curriculum strictly, the ruling group can impose its own education policy by means of a state-controlled construction and implementation process. However, difficult power-sharing mechanisms after conflict and in emergencies rarely invest the government with such exclusive authority.

On the contrary, in most cases where the curriculum is used as the main tool for adjusting the education system to the new conditions, it represents a compromise between the different political factions. Rose and Greeley make the compromise character of the curriculum almost conditional for the success of transformation, stating that it "is important to implement curriculum reform gradually to ensure national consensus is reached."[20] However, the insistence on bringing together former adversaries and on compromising contributes not only to the long duration of the process; it may also have the shortcoming of hiding aspects that could not be agreed upon. In the end, when no compromise can be achieved in matters of crucial importance and curricular tools like syllabuses and textbooks are determined to be the only transmitters of change, the teaching of the subjects and topics at stake and the development of the respective textbooks can be suspended. In this case, what otherwise is used as a provisional, short-time emergency measure to be implemented only in the very beginnings of intervention, turns into a longer-term impediment to change.

This is just such the case in Rwanda. Until now no history textbook could be developed there, since the experts could not succeed in writing a book that integrates the genocide into an accepted overarching interpretation of Rwandan history, from pre-colonial times over colonialism up to the present. Governmental policy emphasizes solidarity and social cohesion so much that differences or internal struggles between the population groups can be referred to only insofar as they have been caused by external (colonial) intervention.

Harvey M. Weinstein, Sarah Warshauer Freedman, and Holly Hughson argue that "since official government policy is to repress Hutu-Tutu difference in favour of a Rwandese identity, the history and traditions of the groups may not be acknowledged. . . . The question is whether the suppression of these identities will result in a unified civic identity or lead to an underground adherence to ethnic difference that ultimately might result in renewed violence."[21]

Moreover in Lebanon, which represents a highly compartmentalized society and educational system, no common history textbook could be produced that met the approval of all religious communities of the country, a requirement of the Ta'if Agreement which ended civil war. Up to now, each community continues to define its own way of how history should be taught to their pupils.[22]

The difficulty of writing a curricular textbook on contentious issues of national identity that could serve as an official guidance for explaining the conflict is shown in exemplary way by Héctor Lindo-Fuentes, who undertook exactly this task after the civil war in El Salvador ended.[23] In contrast to other cases, like Rwanda and Lebanon, he succeeded in finalizing the book and getting approval, but such a book may hide more than it reveals about the roots, causes and agents of the conflict. In search of compromise and legitimacy for the government, the curriculum can often avoid openly addressing the reasons for conflict, social unrest, political fragility, or economic disaster.

More open forms of teaching must be allowed and trained if the possible roots of the conflict situation are to be addressed at all and not neglected.

To avoid deadlocks, a curriculum that addresses contentious issues should offer choices and give opportunities to examine, compare, and discuss different interpretations and methods instead of just replacing one concept or methodology with another. This approach implies a mutual respect and acknowledges differences rather than aiming at constructing a new, albeit superficial, harmony that it is not likely to stand the test of new, or the resurgence of old, conflicts. It encourages teachers' initiatives and takes into account local characteristics. The point of view of the "other" is acknowledged and placed under scrutiny in the same way as one's own view.

Such an approach is more difficult for teachers. They need help to get acquainted with new forms of teaching, with the changing role they have in discussion groups, where they no longer are the sole transmitter of knowledge and ideas, but act as a chairperson and mediator. Drawing on the experience she gained in the reconstruction of the education systems in Burundi and Rwanda, Ana Obura recommended establishing discussion groups and in-service training with teachers as a first step, before moving on to curriculum development.[24] Often, the sequencing is the other way round: it begins with quick emergency measures, such as provisionally screening or updating curricular tools like textbooks and syllabi. As a second step, a longer process of curriculum starts, and, only after this is completed, are considerable resources provided for training the teachers in how to use the new curriculum.

Summarizing the findings, one can conclude that the following factors play a decisive role in coping with the dilemmas of handling curriculum development in emergencies and post-conflict situations:

- The willingness of the ruling powers, the educational professionals and interest groups, parents' associations in particular, to cooperate.
- The ability of teachers to show initiative and develop their own teaching plans and materials when regular teaching devices are scarce or not available.

- The ability of all the main stakeholders to coordinate their activities in order to transmit a clear message to teachers, parents, and students.

In view of the multitude of stakeholders, interests and approaches coordination is of the essence where systematic and concerted action is not possible and, under some conditions, not even advisable.

10 | Attacks on Education

BRENDAN O'MALLEY

In November 2008, two motorcyclists rode up to a group of school-girls and teachers chatting on their way to Mirwais Nika Girls High School in Kandahar, southern Afghanistan, and threw liquid over them. Atifa Biba, fourteen, screamed as she felt and smelled her skin burning. The liquid was battery acid. When one of Atifa's friends tried to wipe the liquid from her face, the assailants threw it over her, too, and then over others. The attack left at least one girl blinded and two permanently disfigured. The attackers were reportedly paid 100,000 Pakistani rupees ($1,190) for each of the fifteen girls they were able to burn.[1]

Sadly, this was just one of a growing number of attacks on students and teachers in Afghanistan, and one of thousands of incidents in countries across the world in recent years in which students, teachers, and academics and other education staff have been kidnapped, imprisoned, beaten, tortured, burned alive, shot, or blown up by rebels, armies, and repressive regimes; or recruited or sexually violated by armed groups or armed forces

Long after the funerals have taken place, the effects on education of such incidents are felt through the loss of teachers and intellectuals; the flight of students and staff; grief and psychological trauma among students and personnel; the damage to buildings, materials, and resources; staffing recruitment difficulties; degradation of the education system; and fear of turning up to class.

When the first global study on this problem, *Education under Attack*,[2] was published by UNESCO in 2007, its finding that the number of reported attacks on education had dramatically increased in the preceding three years was publicized around the world. Some of the worst affected countries were Afghanistan, Colombia, Iraq, Nepal, the Palestinian Autonomous Territories, Thailand, and Zimbabwe. The follow-up study, *Education under Attack 2010*, found that systematic targeting of students, teachers, academics, education staff, and institutions had continued in a greater number of countries since then.

The dramatic intensification of attacks in Afghanistan, India, Pakistan and Thailand; and the sudden explosion of attacks in Georgia and Gaza have been among the most worrying new trends in 2007 to 2009. The failure to eradicate the problem in Colombia, Nepal, and Iraq are matters of grave concern. An often overlooked but significant worry is the number of countries where academics are being targeted. In 2009, Radhika Coomaraswamy, the Special Representative of the Secretary-General for Children and Armed Conflict (SRSG-CAAC), told the Human Rights Council: "The escalation in the number of systematic and deliberate attacks on schoolchildren, teachers and school buildings is alarming, as these attacks not only damage property and cause harm to students and teachers, but they also incite fear and limit access to education services by children. Particularly disturbing is the targeting of girl students and girls' schools such as in Afghanistan, warranting the increased attention of, and action by the international community."[3]

The number of targeted attacks on schools tripled in Afghanistan in 2008 to 670. From 2006 to 2008 there were 1,153 attacks on schools, and hundreds of students, teachers, and education staff were killed. In Pakistan, 173 schools were destroyed or damaged by the Taliban as it took control of Swat District between 2007 and March 2009. By the time the Pakistani army had driven them out three months later, that figure had risen to 356. In Iraq, seventy-one academics, two education officials, and thirty-seven students were killed in assassinations and targeted bombings between 2007 and 2009. In India, buildings at nearly three hundred schools were reportedly blown up by Maoist rebels between 2006 and 2009.

In Thailand, teachers have been picked off for assassination one by one or blown up in security escorts to or from school. In numerous cases they have been shot in front of their classes, and in other incidents they have been beheaded or burned alive in front of school. Ministry and press reports suggest that by July 2009, 119 teachers had been assassinated since the start of the Muslim separatist insurgency in the south in 2004. Hundreds of schools had been firebombed. The number of schools attacked quadrupled between 2006 and 2007 to 164, then fell right back.

In Colombia, 90 teachers were murdered between 2006 and 2008. At least 360 teachers' union members were killed and 342 given death threats in the decade to 2009, according to FECODE, a teachers' union.

The targeting of kindergartens, schools, vocational schools, and universities during Israeli military operations in Gaza at the turn of 2008–2009 left 265 students and teachers dead and nearly three hundred educational buildings damaged or destroyed.[4] Approximately 127 nursery schools, schools, and universities were destroyed or damaged in Georgia in August 2008, most of them in Russian-led military operations, but with 28 in Georgian operations in Tskhinvali, South Ossetia.[5] In Africa the worst-affected countries were the Democratic Republic of the Congo, Somalia, Sudan, and Zimbabwe.

In 2008 an estimated 250,000 child soldiers were involved in conflicts around the world or remained in the ranks of armed groups, security forces or armies.[6] There have been reports of children being voluntarily or forcibly recruited from school, or en route to or from school, in 2006 to 2009 in the following countries: Afghanistan, Burundi, Chad, Colombia, the Democratic Republic of the Congo, India, Iran, Iraq, Myanmar, Nepal, Pakistan, Palestinian Autonomous Territories, the Philippines, Somalia, Sri Lanka, Sudan, Thailand, and Zimbabwe. Fear of recruitment by this method has driven parents in Venezuela, on the Colombian border, and in Sri Lanka to keep their children home from school.[7]

Sexual violence against schoolgirls and women continued as a common characteristic of conflict. It reached endemic proportions in DR Congo in 2006 and continued at seriously high levels between 2007 and 2009, with 2,727 cases of sexual violence against children in Oriental province perpetrated mainly by members of armed groups, but also by soldiers and national police officers in 2008.[8] "Most girls are targeted on their way to or from school, as well as at the market," said Sayo Aoki, former UNICEF education specialist in Goma.

Incidents of sexual violence resulting from abduction or attacks at schools or education facilities, or on the journey to or from them, were reported in the Democratic Republic of the Congo, Haiti, Indonesia, Iraq, Myanmar and the Philippines in 2007 to 2009.[9]

New research by the Institute of International Education, conducted among applicants for its relocation grants for endangered scholars, suggests that the persecution of academics is most common in sub-Saharan Africa, Iraq, Gaza, and Iran.[10]

Mostly, attacks on education occur in conflict-affected countries or under regimes with a poor record on human rights and democratic pluralism. Students, teachers and academics have been either beaten, arrested, tortured, threatened with murder, or shot dead by state forces or state-backed forces in Argentina, Bangladesh, Brazil, China, Colombia, Ethiopia, Honduras, Iran, Myanmar, Nepal, Senegal, Somalia, Sudan, Thailand, Turkey, Zambia, and Zimbabwe in 2007 to 2009.[11]

The Nature of Attacks

The types of education attacks experienced in different locations around the world over the past three years include:

- mass or multiple killings or injuries caused by explosions, rocket and mortar attacks, gunfire, or mass poisoning;
- assassinations or attempted assassinations;
- injury and beatings of targeted individuals;
- abductions, kidnappings, forced disappearance, illegal imprisonment, and torture;
- indiscriminate disproportionate violence and targeted violence against education protestors;
- sexual violence by armed groups, soldiers, or security forces against schoolchildren and teachers;

- child soldier recruitment, forced or unforced, under fifteen years old, and use of child suicide bombers, including abduction and recruitment/recruitment from or on the way to or from school, or recruitment that denies access to education;
- destruction of educational buildings, facilities, resources, and learning materials by remotely detonated explosions, mortar and rocket fire, aerial bombing, burning, looting, and ransacking;
- occupation of educational buildings or facilities by the military, security forces, armed police, or armed groups or their use as a military base;
- verbal or written warnings to stop teaching, close schools or other education institutions, not repair or reopen them, not attend school or college, or face violent retribution;
- official published threats or orders by armed groups, the military, or security forces to carry out any of these attacks.

Two disturbing new tactics are the apparently increasing number of direct attacks on schoolchildren and the mass abduction and indoctrination of children—up to 1,500 kidnapped from schools and madrassas according to the army—in Pakistan to become suicide bombers.

The types of attackers vary in different situations between nonstate and state actors. They include groups of armed civilians, civilians mobilized by armed groups, militants, armed groups, armed criminal gangs, state-backed paramilitaries, state armed forces, state police, and state security forces.

Motives for Attack

Developing a deeper understanding of the motives for attacks on different targets is vital to any attempt to prevent education being targeted in future. Analysis is hampered by the lack of high-quality reporting and monitoring, based on large numbers of face-to-face interviews, in many conflict countries and in situations where the perpetrators are repressive regimes.

From the available information, the motives for attack tend to fall into the following categories:

- attacks on schools or teachers as symbols of imposition of alien culture, philosophy, or ethnic identity;
- attacks on schools, teachers, and students to prevent education of girls;
- attacks on schools, teachers, and students to prevent any education;
- attacks on schools as symbol of government power opposed by rebels;
- attacks on schools, universities, education offices, students, teachers, other staff, and officials to undermine confidence in government control of an area;
- attacks on schools, teachers, and students in revenge for civilian killings;
- attacks on exam halls and exam transport, or ministry or local district offices and officials, to undermine the functioning of education;
- abduction of children and some adults to fill ranks or logistical support of rebel or armed forces or provide forced labor or sexual services;
- abduction for ransom;
- sexual violence by armed groups, soldiers, or security forces as a tactic of war or gender-rights disrespect;
- attacks on students, teachers, teacher trade unionists, and academics for involvement in trade union activity;
- attacks on students and academics to silence political opposition or prevent voicing of alternative views;
- attacks on students and academics to silence human rights campaigns;
- attacks on academics to limit research on sensitive topics;
- occupation of schools for security/military operations by security forces/armed forces/armed groups and attacks on them by rebels for the same reason;

- destruction of schools by invading forces as tactic of defeating the enemy;
- destruction of educational buildings in revenge for or to deter mortar, rocket, or stoning attacks launched from inside them or nearby.

Attackers may have multiple motives. In Thailand, for instance, Islamic separatists in the three southernmost provinces may attack schools and kill teachers because schools are the main symbol in a village of the power of the government they oppose and are easier to attack than other targets. But they may also be targeting them because schools are seen as imposing an alien language of instruction (Thai), religion (Buddhism), and history (Thai national history) as part of a historic policy to assimilate Muslims in a previously autonomous area.

In Afghanistan and Pakistan the gender motive behind many attacks is underlined by formal and written threats from the Taliban. When a former Director of Education in a province with 480 schools in Afghanistan, attended the launch of the *Education under Attack* report in New York in 2007, she brought with her a handful of written death threats received by herself and other officials, head teachers, and teachers. One demanded, "Close schools for girls . . . it is the Islamic duty not to continue with those schools." Others were vaguer about their motives: "I have a request that you stop doing this work. . . . If you continue I will kidnap you, take you in a car and kill you." This could have referred to the work of running girls' schools, or any schools. Similarly one signed by the office of Mullah Omar, the Taliban leader, said: "If continue with schools you have not reason to complain for what happens to you."

But the reality shown by data is that while girls' education is disproportionately targeted, mixed and boys' education is also targeted, suggesting multiple motives for attacks. A CARE/World Bank study found that 40 percent of schools attacked in Afghanistan were girls' schools, 32 percent were mixed schools, and 28 percent were boys' schools.[12] In Swat Territory, Pakistan, government figures show that from 2007 to March 2009, 116 girls' schools were destroyed or damaged, but 56 boys' schools were also targeted.[13]

The raiding of schools in the Democratic Republic of the Congo is heavily driven by the need to recruit personnel or provide shelter for troops. In the Naxalite conflict in India, Maoists are blowing up schools because they allegedly have been used as camps by police and paramilitary police forces for security operations. The Naxalites are also on record for issuing one of the most unusual—and paradoxical, given their record on dynamiting school buildings—threats regarding education. A diktat posted on walls by Maoist guerrillas in Bihar, India, in June 2009 warned that they would punish parents from marginalized communities if they did *not* send their children to school.[14]

Impact on Education

While the physical short-term impacts of attacks on education are obvious but rarely reported, the long-term psychological, financial, qualitative, and ideological effects on the education system and the development of regions and countries are barely examined at all.

The physical effects of the bombing, shelling, and burning of school and university buildings include loss of life, injury, the loss of places in which to learn, of learning materials, equipment, and school furniture, research materials, computers data, management information systems, specialist laboratories, transport vehicles.

The psychological effects of the murder, disappearance, and torture of students, teachers, and other staff are much harder to measure. They can include trauma, fear, insecurity, demotivation, and despondency among students, parents, teachers, academics, support staff, aid workers, trade unionists, and officials managing the system.

The fact that some of the most brutal killings are carried out in front of schoolchildren—teachers shot and set alight in front of school, or a severed head placed on the school gate in Thailand; a raped teacher's body strung up by the feet and hung outside a school for days in Iraq—demonstrates the intention to spread fear, but at what severe psychological impact on those children who witness them?

In all these types of attack the educational impact can go way beyond the number of actual victims and beyond the numbers of people threatened. Public knowledge of repeated attacks and threats of attack gleaned from word of mouth or the media can spread fear throughout schools and universities, which may be the intention of the perpetrators. Children will be afraid to go to school, parents will be afraid to send them, teachers will be afraid to go to work.

As Vernor Muñoz, Special Rapporteur on the Right to Education, said, "These attacks have a terrible physical effect, because they destroy human lives, buildings and spaces for safe learning. But they also have a symbolic effect that is devastating in itself, but which also exacerbates the physical effect. The symbolic effect is based on fear, on the subordination of some persons to others and the elimination of opportunities to live with dignity and freedom. There is also an ideological effect of the attacks against schools, in the removal of a basic human right, the right to education. The attacks on educational institutions, students and teachers mean direct and brutal attacks on the human condition."[15]

Where incidents occur, schools will be closed in their tens or hundreds for anything from a week to months or years in some cases. In Afghanistan, 670 schools remained closed in March 2009, denying education to 170,000 children. The Taliban's warning in Swat District, Pakistan, in December 2008, that classes for girls must be terminated led to the closure of 400 private schools and 500 state schools, depriving 120,000 girls of their right to education. In Thailand, 260 schools in thirteen districts closed after two teachers were shot in front of their classes on June 11, 2007, in Narathiwat Province.[16]

In Iraq, at the height of attacks on education, the entire education system was in danger of collapsing, with attendance at school down from 75 percent in 2005–6 to 30 percent in 2006–7, according to Education International.[17] The assassination of hundreds of academics and the killing of dozens of students in targeted bombings drove university attendance down by 40 percent in some university departments.[18]

In the longer term, governments may be reluctant to repair or reopen schools or invest more resources in them until the threat has passed.

Targeted assassinations, forced disappearances, arbitrary detentions, torture and beatings, and threats of any of those things can have a stark impact on the ability and motivation of teachers, trade unionists, academics, and officials to carry out their job leading to a degrading of education quality in the affected areas.

Governments may struggle to recruit teachers in areas where they are targeted for assassination, for instance, and if the problem continues over a number of years this could have serious consequences for the quality and numbers of teachers recruited.

Attacks on aid workers can lead to the removal of aid workers from that area or even from the whole country, with the loss of expertise and supplies that came with them. In Afghanistan in 2009, UNICEF's work as the lead agency supporting education was inhibited by the fact that UN staff were only authorized to work in half the country for security reasons, so work in other provinces had to be carried out by proxy through local NGOs.

The loss of schooling can drive pupils down an alternative path. For example, in some conflicts, out-of-school children are targets for child-soldier recruiters.

It is commonly argued by proponents of providing education in emergencies that it is vital to establish schooling as quickly as possible among conflict-affected people, because it provides stability, investment in the future and a place where children will be cared for, freeing up parents to set about the task of reconstructing their communities and their livelihoods. But the reverse is also true: if schools are repeatedly attacked and it becomes too dangerous for teachers to turn up to teach or for parents to send students to school, a keystone of stability is removed from the community, increasing the likelihood of displacement and putting children at risk from other dangers. It is in the interests of all concerned with child protection to campaign for protection of schools.

As Nick Burnett, UNESCO's Assistant Director-General, Education, has said: "Clearly we will not reach the Millennium Development Goals and Education for All goals as long as children living in fragile and adverse circumstances are excluded from school or only have access to limited and low-quality learning opportunities. We will not reach these goals when students themselves, their teachers and other education personnel are victims of threats and attacks. "But the issue goes well beyond numbers. It is about how the denial of education in fact perpetuates a cycle of violence and exclusion, how we deprive children, youth and adults of the knowledge, skills and values they need to build a better and more peaceful future."[19]

Attacks on schools, universities, students, teachers, academics, and all other education personnel are not just an attack on civilians and civilian buildings. They are an attack on the right to education, including the right to a good-quality education; an attack on academic freedom, an attack on stability, an attack on development, and an attack on democracy.

We already know that there are strong correlations between fragility and low access and achievement in primary education—40 percent of primary-aged children not in school are found in conflict-affected countries, for instance, and the World Bank has argued that provision of education, particularly for girls, is the single most effective intervention that countries can make to improve human development. The prevention of attacks on education and the strengthening of the right to education at all levels, therefore, can also be viewed as an important contribution to tackling conditions of fragility.

Protection Measures

Protection can be interpreted in different ways. There can be protection provided by armed guards at schools and security escorts on school transport. There can be early warning systems combined with mobilization of the local community to confront attackers or negotiate with their representatives. There are measures that can be taken to reduce the visibility of the target, such as relocating schools in village houses, or providing distance learning. And there are measures to rapidly repair, reconstruct and resupply schools to minimize the impact of attacks and protect the right to education. Aid agencies can also factor in the need for rapid supply of IDPs who are sheltering in schools with plastic sheeting so they can camp elsewhere and allow lessons to continue.

The use of dedicated armed protection has had mixed results against attacks on education. Measures that can be taken include a general increase in force levels in the area, security patrols around schools, the posting of police or armed security guards at schools, and security escorts for teachers and students to and from school.

Whether these are provided depends both on the availability of security manpower and the security chiefs being persuaded of the importance of defending education targets. In least-developed and fragile states, or in areas populated by marginalized groups, the task of persuasion may be harder either because education is less valued or because the ruling elite is less motivated to protect the interests of marginalized groups. Cooperation is needed between ministries of education and security-related ministries on this issue.

In Iraq in 2007, national school-leaving exams were struck by militants who entered exam halls and killed teachers and students. MPs pressed for a different approach in 2008, for cooperation between the army, police, and security and education ministries. The result was that for the first time the exams were moved into university buildings where they could more easily be protected.[20]

In Israel, since the massacre at Mai'a lot school in 1974, Israeli children have lived for more than thirty-five years with armed security in every school, on every field trip, and on every armored school bus.

In southern Thailand the government has provided armed escorts to teachers on their way to and from school, but militants have increasingly turned to using remotely detonated bombs to blow up the teachers and their guards en route.

Community Defense

In the poorest countries, particularly those with poor communications and thinly spread rural populations in challenging terrain, stationing troops or police at schools may not be a viable option due to the cost and manpower required.

Two initiatives in Afghanistan have shed new light on alternative ways to defend schools. The first is the community defense initiative launched in June 2006, which involves mobilizing local communities to deter or resist attacks. School-protection *shura* (councils) were established, where school management committees did not exist. They are to be supported by a national information gathering system to provide early warning of attacks. In addition, local people are encouraged to confront attackers, and in cases in at last eight provinces they rushed out to defend schools.

The CARE study found that in two provinces, Balkh and Khost, 12 percent of people said there were cases of attacks being prevented. In those provinces, several respondents said, communities have negotiated with the potential attackers or gained "permission" to continue teaching. In Herat in western Afghanistan, for instance, a police officer recounted how the police and community collaborated in the aftermath of an attack. The community arranged to meet the alleged attackers, the Taliban, and negotiated a halt to attacks, allowing teachers and students to return to school.[21]

Others measures employed in those provinces included communities banning strangers from entering their village, hiring night guards, or patrolling schools themselves, particularly after the receipt of night letters (threats). Even if they were not able to deter an attack, night guards were able to stop schools and tented classrooms from burning down after attacks. In some cases they or community members fought with attackers.

A stronger message coming from the CARE research is that where there is clear community involvement in the running of community affairs, the schools, or their defense, schools seem less likely to be attacked, and negotiated prevention seem more achievable. For instance, villages with a Community Development Council reported far fewer attacks on schools. The hiring of guards for schools or other defense measures by the community was seen to send a message that the schools are "for the people and not for the government," and so should not be a targeted as a symbol of government. By contrast where schools were built without the community requesting them, or rebuilt by international forces, they were thought more likely to be attacked.

Negotiating the Reopening of Schools

The second initiative was a remarkably successful attempt to negotiate the reopening of schools closed due to the threat of attacks by addressing the feelings shared by the community and the armed opposition, the Taliban, that schools were imposing alien values.

Afghanistan's education minister, Farooq Wardak, when he was appointed in late 2008, decided to address the opposition to government schools directly. For at least three years, the government had worked together with UNICEF and other agencies at local level to persuade local elders, particularly religious leaders, to drum up support for the education of girls and the establishment of community-based schools located inside people's homes with volunteer teachers; there are now eight thousand such schools, and with better systems of training and pay to professionalize the teacher workforce.

Encouraged by the success of this initiative, Wardak asked religious and village leaders to mobilize support among local people for the reopening of schools in areas where they had been kept closed after attacks. He then invited influential people from right across the communities, including supporters of the opposition or antigovernment elements such as the Taliban, for consultations to tease out why schools have remained closed and suggest compromises that would increase a sense of local ownership of schools and their curriculum.

Some objected to using the term school, others feared the curriculum or regulations were anti-Islamic. So the government allowed the word *school* to be changed to *madrassa*. It is also permitted communities to nominate a locally trained teacher of their choice to join the school's staff and ensure that nothing anti-Islamic was taking place. And it challenged people to root through the curriculum or textbooks and, if they found anything anti-Islamic, it would be changed.

Schools have also been told to build prayer breaks into the timetable. "It gives a sense that this is our school, the way we want it. The government is compromising and the communities are becoming motivated," said Fazlul Haque of UNICEF, "They are taking care to ensure schools are opened—and nobody is attacking them."[22]

Within the first three months of 2009, as a result of this initiative, 161 schools reopened, compared to 35 in 2007–8. And in the crucial first month of term, when schools are particularly vulnerable to attack, there were no violent incidents.[23]

Recovery Measures for Schools

Apart from prevention, a second way to minimize the impact of attacks is to support rapid recovery. UNICEF has a longstanding policy of visiting schools within seventy-two hours of their being attacked and providing an emergency education package of tents for classrooms, and boxes of teaching and learning materials within five days. In Afghanistan that job is made harder by the fact that, for their own safety, UN staff can no longer travel in half the country, so local partners have to be used.

In some situations a major obstacle is the continuing use of schools as either bases by security forces or shelters for IDPs and negotiations may be required to free the buildings up for their return to use as schools.

In higher education, several of the international networks that advocate on behalf of threatened scholars also help relocate them to the sanctuary of safe institutions in other countries. But there is a need for earlier intervention, before the point at which academics are under such pressure from death threats or experience of arrest or torture that they have to leave a country.

Negotiating Safe Sanctuaries

The right to education in safety was enshrined in the expanded commentary of the Dakar Framework for Action issued by the World Education Forum. Paragraph 58 declares, "Schools should be respected and protected as sanctuaries and zones of peace." *Education under Attack* reported the suggestion from Save the Children that an international symbol be commissioned to denote that schools are safe sanctuaries and must not be attacked.

Apart from the recent initiative to reopen schools in Afghanistan, the most impressive example of negotiations to accept the idea that schools shall be respected as safe sanctuaries was developed during the conflict in Nepal, where Maoists fought the government for more than ten years (1996–2006). During the fighting, schools were forced to close for political strikes imposed by the Marxists, exams were blocked, school buildings were taken over as military bases, thousands of students and teachers were abducted or recruited as soldiers, and others were deliberately killed.

In 2004, stakeholders urged UNICEF and partner organizations to mount a Schools as Zones of Peace initiative. UNICEF supported World Education and other partners to:

- create a module for negotiating and developing school codes of conduct to safeguard them as zones of peace, in which local community facilitators convene negotiations with Maoists, army, civil-society, and other stakeholder groups to cease targeting schools;
- mobilize civil society to keep the conflict out of schools using local media, which monitored threats to schools through the educational journalists' association;
- provide psychosocial and other support for students affected by conflict;
- provide support and coping skills to teachers;
- teach landmine awareness and protection.[24]

Signs were painted on roofs of schools to prevent the government from bombing them, while messages on posters were designed to raise public awareness of the initiative. The program was piloted in two of the worst affected districts. World Education trained trusted local community members, mostly women, to serve as facilitators.

Melinda Smith, former education cluster lead in Nepal, has written: "The facilitators played the central role in engaging both the government and the Maoists by putting social pressure on each to take part, playing on their desire to be seen in a positive light in both communities. . . . Since the Maoists had brothers, nephews and nieces attending the local schools they were persuaded to support codes of conduct not only to help their relatives, but to be consistent with the Maoists' message of universal free public education."[25]

It was too dangerous for the facilitators to meet with the Maoists directly, so negotiations involved a backdoor shuttle process. A key factor was mobilizing local journalists to act as monitors of any agreed code, which was posted at the front of schools. The codes included clauses such as "No arrest or abduction of any individual within the premises"; "No use of school to camp, never consider school premises as a possible target, no use of school as armed base, no use of school uniforms for camouflaging purposes."[26]

Negotiations in two districts proved successful, with thirty-nine schools in six targeted villages declaring themselves Zones of Peace. A district code of conduct was also negotiated. Local monitoring teams reported that all political activities in schools had ceased as a result of the code. A similar initiative by Save the Children also brought positive results, and the concept was taken up nationally post-conflict, achieving the agreement of codes of conduct in nine districts and 403 schools.[27]

The lessons bear some similarities to the lessons of the negotiations in Afghanistan: conditions for success seemed to include ownership of the process by the community, buy-in from the armed opposition, and local leadership and sense of ownership of schools. The big advantage in Nepal was the Maoists' public commitment to universal education.

Monitoring and Punishment

There are two reasons why we should ratchet up the level of monitoring attacks on education: to help us tackle the problem, and to help us catch the perpetrators.

As Nicholas Burnett of UNESCO said: "Regular monitoring is crucial because silence legitimates an unacceptable situation. In some cases, monitoring can act as a deterrent: in others it can be a tool for negotiation and mediation. Every single attack needs to be documented, investigated and brought to international attention. Without monitoring, we cannot denounce, and without denunciation we cannot act to protect education or end impunity."[28]

There is currently no global monitoring system for attacks on education. But there is a growing international monitoring and reporting mechanism for attacks on schools stemming from international action on children and armed conflict, particularly the recruitment of child soldiers. Until now, it has only applied, however, in countries that are officially listed for recruiting and using child soldiers. By March 2009, the list extended to 14 countries: Afghanistan, Burundi, Chad, Central African Republic, Colombia, Cote d'Ivoire, the Democratic Republic of the Congo, Myanmar, Nepal, Philippines, Somalia, Sri Lanka, Sudan, and Uganda.[29]

The UN Security Council first called for an annual listing of parties who use children under the age of eighteen in armed conflict in 2001 (Resolution 1379). Time-bound action plans for halting recruitment and releasing child soldiers were brought in with Resolution 1539 in 2004. State and non-state actors had to agree such plans with the UN country team or face the possibility of sanctions.

In parallel a Monitoring and Reporting Mechanism (MRM) was established under the Secretary-General's 2005 plan to provide "timely, accurate, reliable and objective" information to the Security Council on six grave violations including "attacks on schools and hospitals." For the first time there was an attempt to put in place a systematic monitoring and reporting mechanism, with country teams putting together MRM task forces in each listed state.

A Working Group of the Security Council was set up, of which a subgroup meets every two months to consider reports provided by the MRM task forces, usually focusing on two conflict situations at a time, and makes recommendations to the parties to the conflict, governments, and the UN. The reports contain lists of all the violations, what action has been taken, what actions could not be taken, and recommendations. The subgroup debates them and makes recommendations on them to the Secretary-General concerning instructions to be given to parties to conflict.

The MRM task force in each country is a coalition of willing partners. The country team is chaired by a representative of the Secretary-General or the UN's Resident Coordinator, with a vice chair usually from UNICEF, or otherwise from OHCHR, UNHCR, DPKO, or the ILO. The task forces are supported by the work of the Office of the Special Representative for Children and Armed Conflict, which helps country teams design and implement action plans. They monitor incidents related to the six gravest violations against children:

- Killing or maiming of children
- Recruitment or use of children as soldiers
- Attacks against schools or hospitals
- Denial of humanitarian access for children
- Abduction of children
- Rape and other grave sexual abuses of children

However, by August 2009, the action plans remained limited to addressing the child-soldier recruitment violation. These have yielded some significant successes, as Human Rights Watch has reported.[30]

- Five parties to the armed conflict in Cote d'Ivoire ended their use of child soldiers after agreeing to action plans to end the practice.
- The government of Uganda has been "delisted" from the Secretary-General's list of violators after removing children from the Uganda People's Defense Forces and local defense units associated with the government.
- Several non-state-armed groups in Burma (Myanmar) have signed voluntary "deeds of commitment" agreeing to end their use of child soldiers, and submitting to independent verification.
- As a follow-up to the Security Council working group conclusions, the SRSG on Children and Armed Conflict has secured commitments from parties to armed conflict in the Philippines, Nepal, Chad, Central Africa Republic, and elsewhere to end the recruitment and use of child soldiers and/or to release children from their forces.

- Some actors have agreed to action plans to demobilize child soldiers, although they have yet to be implemented fully.

By contrast the coverage of sexual violence and attacks on education in the Secretary-General's reports has been minimal, and attacks on education have been barely mentioned—and the proportion of recommendations on these issues reflects this oversight. Since the beginning of 2008, nineteen reports from the Secretary-General and the Working Group have included 141 separate recommendations related to the recruitment and use of child soldiers, but only eighteen recommendations on sexual violence and only six related to attacks on education, according to Human Rights Watch.[31]

Attacks on education have been neglected. The reporting on sexual violence is being addressed by the introduction of sexual violence and killing and maiming as additional triggers for the MRM but so far, attacks on education have been given short shrift, except in Afghanistan.

The interesting finding of *Education under Attack* 2010 is that this is not due the lack of a wide definition of attacks on education in the requests for data. OSRCAAC explicitly requests information not just on attacks on education buildings, but also on the killing and injury of teachers, students, academics, and other education staff. The Security Council Steering Group instructions to task forces also tell them to include information on occupation or forced closure of schools, and assaults against school personnel.

Alec Wargo, program officer at OSR/CAAC, believes that the reason for the lack of information on attacks is the lack of involvement of education-oriented UN organizations and NGOs in the task forces—although UNICEF might strongly disagree on that point—or community of practice of reporting on education attacks. "It's a missed opportunity for them and for the children that they're trying to protect," he says. "If you're talking about accountability and attacks on schools, it's one thing that can bring attention to it which brings funding to [protection] programs but also coalesces those actors in the country task forces around thinking about how do we deal with accountability or how do we deal with trying to bring an end to these violations."

One way to rectify this problem, however, is to make attacks on schools a trigger for listing countries and issuing time-bound action plans. The UN Secretary-General has always asked for all six violations to be listing criteria, so the opportunity is there if organizations concerned about protecting education commit themselves to support monitoring, provide information, and push hard on this issue. Looking beyond the MRM, there is a need to develop a truly systematic global monitoring system for attacks on education, which the MRM could feed into, but which would cover all sectors of education, not just education for children.

Ending Impunity

The problem of lack of prosecution of those responsible for attacks on education has vexed the international community since the publication of *Education under Attack* (2007). There were widespread calls at UNESCO's 2008 conference on protecting Iraqi academics for investigations of those responsible for killing hundreds of university professors and intellectuals. At the UN General Assembly debate in 2009, there was an impassioned plea by H.H. Sheikha Moza Bint Nasser Al Missned of Qatar, to ensure that the perpetrators of attacks on education are punished.

A key question raised in the ensuing debate was what more needs to be done to ensure that perpetrators are actually investigated, tried and made to pay a heavy price. Was the apparent widespread lack of progress on this issue due to gaps in the coverage of attacks on education by international law or the monitoring processes that can inform investigators? Or what could be done to ensure laws that did apply were used to investigate such attacks?

The advice of experts on international humanitarian law is quite clear on this point: attacks on education are already adequately covered by existing laws and conventions, even if there is not as much visibility in the wording as, for instance, there is for attacks on hospitals. There is, however, a strong argument that the conversion of education buildings to use for military purposes should be banned, as it is for hospitals and churches, because conversion of one school can increase the risk of many schools being considered legitimate targets.[32] But for most education attacks the problem is not whether the attacks are covered by law, but creating awareness that they are covered and building up pressure for the law to be applied.

A first step might be to support the training of soldiers and journalists in humanitarian laws applying to attacks on education. Training soldiers and, if at all possible, members of armed groups would encourage restraint and the training of journalists would encourage media reporting, increasing the laws' deterrent effect.

Better monitoring and media reporting would pave the way for more investigations. The onus is on governments to criminalize attacks on education. Radhika Coomaraswamy, the SRSG/CAAC, says that ensuring national legislation relating to genocide, crimes against humanity, and war crimes makes these crimes punishable under national law with the same scope and definition as the Rome Statute would be an important step toward addressing the prevailing culture of impunity on all grave violations against children.

This argument is important due to the complementary nature of the International Criminal Court, which the Rome Statute established. The ICC is an important vehicle for encouraging prosecutions for attacks on education for two reasons.

First, the Rome Statute classifies three child-specific provisions, originally proposed by the OSR/CAAC, as war crimes—the conscription, enlistment or use in hostilities of children under fifteen; grave acts of sexual violence; and attacks on hospitals and schools—and it is already carrying out investigations concerning such crimes.

Currently, the ICC's Office of the Prosecutor is conducting investigations and prosecutions in eight cases, in four conflicts (the Democratic Republic of Congo, Uganda, Sudan, and the Central African Republic), involving fourteen individuals. Six of those cases contain references to crimes that can be described as attacks on education as defined in Education under Attack 2010. Additionally the close relationship of the OSR/CAAC to the courts encourages sharing of information from the MRM task forces on these particular violations, which may encourage further investigations.

Second, although under the principle of complementarity the ICC tries cases only when states are unwilling or unable to do so in line with their responsibility to prevent and punish atrocities, the Office of the Prosecutor takes a proactive stance against impunity, using its influence within national and international networks to encourage and provide support for genuine national proceedings where possible.

Lothar Krappman, a member of the UN Committee on the Rights of the Child, has said, "This vast demolition of education facilities is not only a side effect of blind military action. Let me ask, whether these attacks on schools, teachers and children really get the attention these crimes deserve. States should criminalise attacks on schools as war crimes in accordance with article 8(2) (b) (ix) of the Rome Statute of the ICC and prosecute offenders accordingly."[33]

A useful objective for those concerned with attacks on education might be to encourage investigations in the most high-profile situations where attacks on schools and teachers are well publicized and the perpetrators have made their intention to attack schools public in published threats and orders. A number of situations spring to mind: the targeting of schools in Afghanistan and Pakistan by the Taliban and in India by the Maoists. In Afghanistan there are numerous examples of written threats, some of them in the name of Taliban leader Mullah Omar and many in the name of the Taliban. In Pakistan the threat to girls' schools to close down or face the consequences was made very publicly on behalf of local Taliban leader Maulana Fazlullah and was later reviewed and modified at a meeting chaired by him to apply to schools teaching girls above grade four. In India the Naxalites have made public statements, particularly in Chhattisgarh State, indicating that they have blown up school buildings in retaliation for their previously being used as camps for security forces, and in many cases they have blown schools up long after they were no longer being used as camps. The scale of attacks in Pakistan and Afghanistan and the deterrent message it would send because those conflicts—and attacks on education in those conflicts—are more widely reported, would be a good place to start, though, of course, the alleged perpetrators would have to be caught before they could be tried.

At the same time, states and parties to conflict should be encouraged to take positive steps to uphold the right to education and the right to enjoy education in safety as implied by international law. To this end, the collaborative development of internationally endorsed guidelines for protecting education systems during war and insecurity would be a positive step forward.

In conclusion, where the actors are willing, there may be scope for negotiating respect for schools and other education institutions as safe sanctuaries. Mobilizing communities to defend education may be an important step toward that. But where the actors will not desist, international law must be employed to punish the perpetrators. The human rights instruments already exist to deal with this problem; it is a question of gathering the evidence and putting pressure on international and national courts to investigate more cases. The starting point should be a relentless international campaign of awareness and advocacy to ensure that education attacks, including attacks on higher education, are effectively monitored and investigated and addressed through the courts.

This chapter is based on the findings of Education under Attack 2010 *(Paris: UNESCO, 2010), written by Brendan O'Malley and commissioned by Mark Richmond of UNESCO.*

11 | Minimum Standards, Maximum Results

ALLISON ANDERSON AND JENNIFER HOFMANN

Education is critical for all children, but it is especially urgent for the tens of millions of children affected by emergencies, whether caused by man-made or natural disasters. Armed conflict and natural disasters can significantly damage educational efforts and deny learners the transformative effects of quality education. Education is one of the principal losses in emergency situations: lack of education, too often stretching into the post-conflict phase, endangers well-being and survival.

At the same time, in emergency situations, quality education can play a crucial role in helping children and adults cope with their situation by learning knowledge and skills for survival and regaining normalcy in their lives. Indeed, education can be both *life-sustaining* and *life-saving*, providing physical, psychosocial, and cognitive protection to both students and teachers.

But how does one define *quality* education, who may judge if interventions are of quality or are not, and how do you develop, implement, monitor, and evaluate quality education programs in such challenging circumstances where multiple education providers—the government, NGOs, UN agencies, and private businesses—are active? By the early 2000s, as awareness had increased about the need for nonformal and formal education programs in emergency situations, this issue of how to ensure quality and accountability had come to the fore.

The Inter-Agency Network for Education in Emergencies

At the 2000 World Education Forum in Dakar, delegates recognized that the goal of Education for All (EFA) by 2015 would not be met unless special attention was paid to the education of those affected by crisis. The Inter-Agency Network for Education in Emergencies (INEE) emerged over the next year as a result of these deliberations. INEE is an open global network of members working together within a humanitarian and development framework to ensure all persons the right to quality and safe education in emergencies and post-crisis recovery. Since its inception in 2000, INEE's membership has grown to more than 3,500 members—practitioners, students, teachers and staff from UN agencies, non-governmental organizations, donors, governments, and universities—in 115 countries collaboratively working to share knowledge, develop resources, and inform policy through consensus-driven advocacy.

The members of INEE work together to ensure that:

- All people affected by crisis and instability have access to quality, relevant and safe education opportunities.
- Education services are integrated into all emergency interventions as an essential life-saving and life-sustaining component of humanitarian response.
- Governments and donors provide sustainable funding and develop holistic policies to ensure education preparedness, crisis prevention, mitigation, response, and recovery.
- All education programs responding to emergencies, chronic crises, and reconstruction are consistent with the INEE Minimum Standards and accountable for quality and results.

INEE improves communication and coordination among its member agencies by cultivating and facilitating constructive and collaborative relationships, creating opportunities for information sharing, resource development and capacity building. The power of INEE lies in the fact that lessons learned, good practices, and tools are shared across the network—across geographic and organizational boundaries—and adapted and used to ensure quality, relevant programming around the world. Such a collaborative approach allows for knowledge gained from a Save the Children program in Sudan, an International Rescue Committee (IRC) program in Afghanistan or a UNICEF program in Nepal to be shared and adapted by other agencies responding to immediate needs, education and psychosocial well-being, from the chronic crises in northern Uganda and the Democratic Republic of the Congo to earthquake-affected Pakistan. In these and in hundreds of other contexts and countries, INEE members share and build upon the fields' collective work and knowledge to create better quality and safer education opportunities in emergencies.

The Development of Global Standards for Access, Quality, and Accountability

In March 2002, INEE members gathered in Paris to share common insights and challenges in education in emergency programming and to chart the way forward for the network. At this forum, humanitarian agencies such as CARE, the IRC, Save the Children, the Norwegian Refugee Council, UNHCR, and UNICEF, which had all carried out emergency education programs for children since the 1990s, shared lessons about the life-sustaining and life-saving nature of quality education. They also shared their frustration with the lack of coordination of these efforts, limited funding, the absence of accepted good practice on which to base their interventions, and the need to link improved quality and accountability to advocacy.

INEE members looked at the Sphere Project as a model through which to mainstream education into humanitarian response, enhance levels of quality, access and accountability within emergency education programming and secure increased funding. The Sphere Project was created to improve the quality and accountability of disaster response. It has argued for the universal right of all disaster-affected people to humanitarian assistance. It has achieved NGO agreement on core principles and actions and collected minimum programming standards for disasters from past lessons and experience. The Sphere Minimum Standards represent consensus on key technical indicators and guidance for: water supply, sanitation, and hygiene promotion; food security, nutrition, and food aid; shelter, settlement, and non-food items; and health services. However, the Sphere handbook, comprising the Humanitarian Charter and Minimum Standards, does not address education services.

INEE members believed that it was necessary develop a tool to guide effective action to meet the education rights of affected populations that could also be utilized to promote education as a key pillar of emergency response. Therefore, learning from the example of the Sphere Project, education stakeholders utilized the open, collaborative nature of the INEE network as a platform upon which to develop common, global standards. A Working Group on Minimum Standards for Education in Emergencies was constituted in 2003 within INEE to facilitate the development of global minimum standards for education in emergencies through to early recovery contexts. Over the next year and a half, the Working Group facilitated a global consultative process, involving more than 2,250 individuals from more than fifty countries, including students, teachers, staff of NGOs, UN agencies, donors, governments, and academics. The minimum standards were developed, debated, and agreed upon through a participatory process of:

- Online consultation inputs via the INEE listserv
- Community-level, national, subregional, and regional consultations
- A peer review process

The handbook *Minimum Standards for Education in Emergencies, Chronic Crises and Early Reconstruction* (INEE Minimum Standards) was launched at INEE's Second Global Consultation on Education in Emergencies and Early Recovery, in Cape Town, South Africa in December 2004. The handbook was well received by delegates who judged the consultative process for developing the standards as significant as the product itself.

Content of the INEE Minimum Standards

The INEE Minimum Standards are the first global tool to define a minimum level of educational quality in order to provide assistance that reflects and reinforces the right to life with dignity. The standards, indicators, and guidance notes within the handbook are built upon the foundations of the Convention on the Rights of the Child, the Dakar Education for All framework, the UN Millennium Development Goals, and the Sphere Project's Humanitarian Charter. In addition to reflecting these international rights and commitments, the standards are an expression of consensus on good practices and lessons learned across the field of education and protection in emergencies and postcrisis situations. They are designed for use in risk reduction, preparedness, and response activities and in formulating policy in a wide range of emergency and recovery situations, including disasters, armed conflicts, and complex emergencies.

The minimum standards offer guidance on how to overcome the barriers to quality, safe education in crises settings. They are organized around five main categories:

- Minimum Standards Common to All Categories: focuses on the essential areas of community participation and utilizing local resources when applying the standards in this handbook, as well as ensuring that emergency education responses are based on an initial assessment that is followed by an appropriate response and continued monitoring and evaluation.
- Access and Learning Environment: focuses on partnerships to promote access to learning opportunities as well as intersectoral linkages with, for example, health, water and sanitation, food aid, nutrition, and shelter, to enhance security and physical, cognitive, and psychological well-being.
- Teaching and Learning: focuses on critical elements that promote effective teaching and learning: curriculum, teacher training, learner-centered instruction, and assessment.

- Teachers and other Education Personnel: focuses on the administration and management of human resources in the field of education, including recruitment and selection, conditions of service, and supervision and support.
- Education Policy and Coordination: focuses on policy formulation and enactment, planning and implementation, and coordination.

In addition, the critical issues of gender, participation, HIV/AIDS, disability, vulnerability, and human and children's rights are cross-cutting and thus have been integrated into relevant standards throughout the handbook.

Implementing the INEE Minimum Standards

The INEE Minimum Standards present a global framework for coordinated action to enhance the quality of educational preparedness and response in crisis contexts, increase access to safe and relevant learning opportunities, and ensure accountability in providing these services. They provide good practices and concrete guidance to governments and humanitarian workers to enhance the resilience of education systems and can be used for sector planning. Since its launch at the end of 2004, the INEE Minimum Standards handbook has been translated into more than fifteen languages and is currently used in more than eighty countries around the world to improve program and policy planning, assessment, design, implementation, monitoring, and evaluation as well as advocacy and preparedness in order to reach the Education for All goals.

Users relate that the INEE Minimum Standards provide a common language, facilitating the development of shared visions between different stakeholders, including members of affected communities, humanitarian agency staff and governments. They provide an effective training and capacity-building tool: more than three hundred educational, protection and emergency trainers have been trained on the standards and are training hundreds of others through a cascade training model. The standards are also used to promote holistic thinking and response and to frame and foster inter- and intra-agency policy dialogue, coordination, advocacy, and action for the provision of quality education in emergencies, chronic crises, and early reconstruction.

Evaluation of the INEE Minimum Standards has been ongoing since their launch. Between 2005 and 2007, INEE and key partners conducted a global survey, collecting and analyzing nearly two hundred responses from people working in ninety-five countries on the implementation, institutionalization and impact of the INEE Minimum Standards. In addition, evaluations have been undertaken in Pakistan, in Darfur, and in Uganda to deepen the understanding of the awareness, utilization, institutionalization, and impact of the INEE Minimum Standards.

Key findings from the global survey revealed that the contexts in which the standards are being used are diverse, ranging from situations of conflict to natural disasters and from acute emergencies to postcrisis/reconstruction contexts and preparedness. However, in all cases, the evaluations highlighted the need for more awareness and training on the standards. At the same time, they confirm that those aware of the INEE Minimum Standards believe that the standards positively support education programming and improve coordination efforts. The most frequently reported use of the standards has been for community participation and for assessment, project design, monitoring and evaluation, advocacy, and preparedness planning. In the global evaluation, more than 80 percent of respondents used one or more of the cross-cutting issues in their projects, most commonly gender issues. Nearly 70 percent of respondents incorporate gender issues into their projects, while approximately 40 percent of respondents have incorporated HIV/AIDS and/or special education as cross-cutting issues.

Also in the global survey, over three quarters of the respondents indicated that education in emergencies, chronic crises, and early reconstruction has been incorporated into their institutions, and almost 20 percent of respondents indicated that development of the INEE Minimum Standards and training in their use have been important factors in their institution's decision to prioritize education in emergencies. Overall, 64 percent of respondents indicated that their organizations have committed to using the INEE Minimum Standards, especially respondents from international NGOs and UN agencies.

While assessing impact is difficult, almost a third of respondents in the global survey felt that the use of the INEE Minimum Standards has led to achievements in project outcomes or improvements in the quality of educational services provided in their projects. Overall, it appears that the organization and focus of the INEE Minimum Standards' targeted approach allow practitioners to use those standards that most directly apply to their work. As a result, respondents indicated that they were able to better motivate the community, better advocate for needed facilities, and more effectively train teachers. Specific examples of what respondents said about how the standards have contributed to improved quality are categorized and listed herein.

The INEE Minimum Standards will not solve all problems; however, they do offer a tool for humanitarian agencies, governments and communities to enhance the effectiveness and the quality of educational assistance, and thus to make a significant difference in the lives of people affected by crises. For example, users of the INEE Minimum Standards have reported that using the INEE Minimum Standards helps them better meet the rights of populations affected by crises, better ensure a minimum level of quality and accountability within their interventions, and better promote education as a core part of humanitarian response.

Achievements or Improvements Associated with the Use of the INEE Minimum Standards

Improved Community Participation and Coordination

- Greater effort and results have been made toward enhancement of community participation and ownership as well as toward more effective coordination among stakeholders in education projects.
- The targeted communities felt more responsible for addressing the educational needs of their children and were more involved, positive and supporting.

- The students have set up a students' council to enhance participation of beneficiaries, and this has resulted in ownership of the language classes.
- The community has been assisting in promotion of education, especially for girls.

Increased Attention to Issues Detailed in the Analysis Standards (Assessment, Response, Monitoring, Evaluation)

- The Minimum Standards helped us to make the process of monitoring and evaluation more advanced and effective. The process of planning has also improved.
- The level of staff awareness about the different dimensions of a holistic and adequate educational program for displaced youth has grown. Improvement has followed at the level of project design and monitoring activities.
- All the projects are analyzed by the staff members trained in minimum standards and necessary feedback is provided to the project implementation staff.
- During the Lebanon crisis last summer, our planning was directly influenced by the Minimum Standards and led to partners engaging in activities that allowed for the timely return to school for children in combat areas and their receipt of psychological support when they arrived.

Increased Attention to Advocacy and Capacity Building

- Advocating for education to be accessible to children from ethnic minority groups has grown, as has providing support to ethnic minority children and families to attend school and remain in attendance.
- "My project is . . . a global research and capacity-building project, with advocacy for education in emergencies and reconstruction an important activity. The Minimum Standards are priceless for that. They are also a great focus for networking activity that is vital to my work."

Increased Attention to Specific Issues or Standard Categories

- The standards have improved the quality of lessons, presentation, and classroom management by teachers
- More children are enrolled in schools near their homes. Funding has been obtained for a project to improve the school environment such as access to water and latrines.
- The use of the standards has helped determine clear indicators for quality of education, for the training of trainers (teachers' code of conduct), and for curriculum implementation and classroom organization.
- Special-needs children have been enrolled in school, girls are treated equally in classrooms, and teachers/head teachers and community are more aware about these issues.
- Understanding of methodology and child-centered learning has improved.

A Tool to Meet the Education Rights of Populations Affected by Crisis

The right to free and compulsory primary education without discrimination is enshrined in international law. Educational rights have been further elaborated to address issues of quality and equity, with some agreements directly addressing provision for refugees and children affected by armed conflict. During the development of the INEE Minimum Standards, INEE members reviewed and incorporated the provisions codified in the following most relevant global-rights instruments:

- The 1948 Universal Declaration of Human Rights
  - Article 26 outlines the right to free and compulsory education at the elementary level and urges that professional and technical education be made available. The declaration states that education should work to strengthen respect for human rights and promote peace. Parents have the right to choose the kind of education provided to their child.

- The 1951 Convention Relating to the Status of Refugees
  - Refugee children are guaranteed the right to elementary education in Article 22, which states they should be accorded the same opportunities as nationals from the host country. Beyond primary school, refugee children are treated as other aliens, allowing for the recognition of foreign school certificates/awarding of scholarships.
- The 1966 Covenant on Economic, Social, and Cultural Rights
  - The right to free and compulsory education at the primary level and accessible secondary-level education is laid out in Article 13. The covenant goes on to call for basic education to be made available to those who have not received or completed primary education. Emphasis is placed on improving conditions/teaching standards.
- The 1989 Convention of the Rights of the Child
  - Article 28 calls for states to make primary education compulsory and free to all, and to encourage the development of accessible secondary and other forms of education. Quality and relevance is detailed in Article 29, which mandates an education that builds on a child's potential and supports their cultural identity. Psychosocial support and enriched curriculum for conflict-affected children are both emphasized in this article. Article 2 outlines the principle of nondiscrimination, including access for children with disabilities, gender equity, and the protection of linguistic and cultural rights of ethnic minority communities. Article 31 protects a child's right to recreation and culture.
- The 1990 World Declaration on Education for All
  - In 1990, at a global meeting in Jomtien, Thailand, the governments of the world committed to ensuring basic education for all. Ten years later, at the Dakar World Education Forum, governments and agencies identified humanitarian emergencies as a major obstacle toward achieving the goals of Education for All (EFA). Within the Dakar Framework of Action, a call was made for active commitment to remove disparities in access for underserved groups, notably girls, working children, refugees,

those displaced by war and disaster, and children with disabilities.

- The Geneva Conventions
  - For situations of armed conflict, the Geneva Conventions lay out particular humanitarian protections for people—including children—who are not taking part in hostilities. In times of hostility, states are responsible for ensuring the provision of education for orphaned or unaccompanied children. In situations of military occupation, the occupying power must facilitate institutions "devoted to the care and education of children." Schools and other buildings used for civil purposes are guaranteed protection from military attacks.

Because the INEE Minimum Standards are based on rights, they are qualitative and aspirational in nature. In some instances, their immediate realization may not be possible. When this is the case, it is critical to reflect upon and understand the gap between the standards and indicators listed in the handbook and the ones reached in actual practice. One must also identify the reasons for the gap and what needs to be changed in order to realize the standards. Once that is known, program and policy strategies can be developed and advocacy can be undertaken to reduce the gap.

Given their legal foundations, the INEE Minimum Standards reinforce the responsibility of duty bearers to comply with their human rights obligations regarding education. For example, the Brazilian Report on the Right to Education includes a recommendation to use and contextualize the INEE Minimum Standards to address education challenges in Brazilian favelas. The recommendation suggests using the INEE Minimum Standards particularly in developing and implementing an action plan to respond to education needs and rights within favelas in Rio de Janeiro affected by conflicts between security forces and drug traffickers.

Many education ministries have developed national education standards, and INEE acknowledges the leading role of national governments in defining education policies and ensuring the provision of basic educational services to all children living in relevant countries, including refugees, internally displaced people, and minorities. In situations where there are national standards, it is important to analyze the differences between those standards and the INEE Minimum Standards, performing a gap analysis in scope, intent and content. The evaluation of the INEE Minimum Standards conducted in Uganda in April 2008 revealed that the Ministry of Education was initially reluctant to engaging with the INEE Minimum Standards because it already had standards developed by the Ugandan Education Standards Agency. However, a joint analysis performed with the ministry highlighted that not only are the INEE Minimum Standards not incompatible or competing with the national education standards but also that, instead, they are a useful tool to complement and help reach national standards by providing strategies for their implementation. Similarly, in Afghanistan the Ministry of Education has its own education policies on education, including one on community-based education that was developed in August 2006 and that clearly defines what community-based education is. The contextualized standards based on the INEE Minimum Standards developed by the Partnership for Advancing Community Education in Afghanistan (PACE-A), a consortium of international organizations established in April 2006, serve to complement and operationalize the ministry's policy. In Lebanon, the United Nations Relief and Works Agency (UNRWA), which operates one of the largest school systems in the Middle East and has been the main provider of basic education to Palestinian refugees for nearly five decades, recently opened up to collaborating with international agencies from the United Nations and the NGO sector and attended a training workshop on the INEE Minimum Standards cohosted by UNESCO and the Norwegian Refugee Council in August 2009 in Beirut to support access to quality education for all children, including Palestinian and Iraqi refugees.

A Tool for Quality and Accountability Within Humanitarian Response

In his 2008 Report on the Right to Education focusing on education in emergencies, Special Rapporteur to the Secretary-General of the

United Nations Vernor Muñoz recommended that intergovernmental organizations and NGOs should "guarantee that educational responses to emergencies are in line with the INEE Minimum Standards." The INEE Minimum Standards Handbook is not a legally binding document, and therefore a commitment to respecting the standards does not imply a moral or financial obligation to meeting them. However, it is highly recommended to use them because they contribute to good practice, quality and accountability within education programming and planning. An INEE member in Iraq noted, "Even if every standard was not met, due to the reality on the ground, the INEE Minimum Standards were very useful as a supporting tool during the design and implementation of the programs. The Minimum Standards inspired a more participatory approach to our work in Iraq, and helped us come closer to ensuring a quality education for all—both boys and girls—despite many obstacles." In emergency situations such as in Indonesia, the Democratic Republic of Congo, or Pakistan where a flurry of organizations suddenly opens up education programs of sometimes varying quality, the INEE Minimum Standards can help identify who is respecting global good practices and who is not.

In fact, donor agencies themselves are becoming more and more aware of the importance of using the INEE Minimum Standards to verify that the education programs they support are based on widely accepted global good practices and rights. While CIDA policy documents do not to date explicitly mention the INEE Minimum Standards, CIDA officers and education specialists who review proposals from partner NGOs and other civil society organizations do advise them to refer to the INEE Minimum Standards and to use them in their overall education programming as a "reference document" and an "analytical framework." The case study developed by INEE on CIDA's experience revealed that there have been a number of positive discussions between CIDA officers and implementing partners that led to the integration of the INEE Minimum Standards in education programming, particularly with regard to monitoring and evaluation tools.

Within the Norwegian Agency for Development Cooperation (NORAD), the Education and Research Department does not itself extend financial grants but provides technical advice to its Civil Society

Department, which allocates grants for projects and programs on education in emergencies. When asked for advice by the Ministry of Foreign Affairs and Norwegian Embassies, the Education and Research Department recommends that organizations applying for financial support describe their use of the INEE Minimum Standards.

Along with institutionalization and interagency coordination, donor engagement on the INEE Minimum Standards is key to ensuring the quality and accountability of education interventions in humanitarian and postcrisis recovery settings. CIDA and the government of Norway are two out of the five governments that have explicit policies on education in emergencies, so it is not surprising to see that they are leading the utilization of the INEE Minimum Standards within the donor realm—a responsible move that, it is hoped, other agencies will follow.

A Tool to Promote Education as a Key Humanitarian Response

The Inter-Agency Standing Committee (IASC), which coordinates global humanitarian policy, established an Education Cluster at the end of 2006. The inclusion of education within the IASC cluster initiative is a significant achievement, since it indicates recognition by the international community of the critical role that education plays in humanitarian response. Jointly led by UNICEF and the Save the Children Alliance, the cluster represents a groundbreaking commitment to response predictability, preparedness, policy, and coordination within the field of education in emergencies. The cluster has adopted the INEE Minimum Standards as a guiding common framework, and as such, the implementation and institutionalization of the INEE Minimum Standards will be carried out in partnership with the IASC Education Cluster, providing an opportunity to bring the standards to a wider audience in a more systematic way. INEE's ongoing involvement in and support for the cluster will guarantee a scaffold of technical knowledge, tools, and information-gathering and information-sharing practices that will greatly enhance the cluster's ability to identify and address capacity gaps and bring actors together at country and global levels to ensure a more predictable, timely, and effective education response. The work of the IASC Education Cluster, and the use of the INEE Minimum Standards within it, will serve to strengthen capacity and preparedness of humanitarian personnel and government authorities to plan, coordinate, and manage quality educational programs in emergencies.

Already in Uganda, Madagascar, the Philippines, and the Democratic Republic of Congo, the education cluster has adopted the INEE Minimum Standards as a framework for emergency preparedness and response plans. This means that the government, UN agencies, international agencies, and local NGOs come together around the common language and guidance of the INEE Minimum Standards to plan holistically for access to quality response to education, even ahead of likely emergencies to occur. It is a considerable shift from the isolated, uncoordinated responses that characterized humanitarian action not so long ago.

Moreover, in October 2008, the Sphere Project and INEE announced the signature of a "Companionship Agreement" between the Sphere Project, *Humanitarian Charter and Minimum Standards in Disaster Response* (the Sphere Handbook) and the INEE Minimum Standards. By this agreement, the Sphere Project acknowledges the quality of the INEE Minimum Standards, and of the broad consultative process that led to their development. As such, the Sphere Project recommends that the INEE Minimum Standards be used as companion and complementing standards to the Sphere Handbook.

This companionship agreement is an important achievement toward one of the main objectives of the network itself: that education services be integrated into all humanitarian response. For an effective education response that addresses children's holistic needs, coordination, and close collaboration between education and other sectors is essential—particularly water and sanitation, shelter, camp management, health and hygiene, protection, food aid, and nutrition. An inter-sectoral approach to education is even more vital in emergency contexts than in normal situations. Education in safe spaces provided for children in an emergency context, offers a means of providing a sense of normality, psychosocial support, and protection against harm, as well as a place for delivery of other vital services. The use of the INEE Minimum Standards as a companion to the Sphere Handbook will help to ensure that these crucial linkages are made at the outset of an emergency—through multi-sectoral needs assessments, followed by joint planning and holistic response. Used together, the Sphere and the INEE Minimum Standards Handbooks will improve the quality of assistance provided to people affected by crisis, and enhance the accountability of disaster preparedness and response.

Next Steps: Ensuring Maximum Results

INEE Minimum Standards Update

Since the launch of the INEE Minimum Standards, significant changes have taken place in the field of education in emergencies. First, a number of issues have evolved or emerged, notably around climate change, disaster risk reduction, and conflict mitigation. Second, as mentioned before, the Humanitarian Reform process was launched in 2005 and aims "to enhance humanitarian response capacity, predictability, accountability and partnership," and the handbook needs to explain and take into account the IASC Education Cluster. Third, the companionship between the Sphere Project Humanitarian Charter and Minimum Standards in Disaster Response and INEE Minimum Standards calls for a better representation of the strong linkages between the two sets of standards. Based on recommendations received from the INEE Minimum Standards feedback forms (collected from 2005 to 2008) and verbal recommendations received from members at trainings workshops and various meetings over the past four years, three key areas have emerged as priorities in carrying out the update: reflecting developments in the field of education in emergencies and post-crisis recovery; being more user-friendly; and incorporating the experiences of INEE members using it so far.

The purpose of the update process is therefore not to change the qualitative standards, nor to overhaul the INEE Minimum Standards Handbook. Rather, it is to simplify its format and language while strengthening key issues as needed (for example, DRR, HIV/AIDS, curriculum, and so forth) and making qualitative indicators and guidance notes more specific and quantifiable. The 2010 edition of the INEE Minimum Standards, which is expected to be launch in the spring, will also address inconsistencies, faults and omissions from the 2004 edition.

INEE Minimum Standards Application

INEE will continue to provide full support to its members with the application of the INEE Minimum Standards, including developing tools and case studies. In particular, INEE will strengthen its collaboration with:

- Ministries of Education, to use the INEE Minimum Standards as a policy and planning tool as well as complementing standards to already existing national education standards.
- Donor agencies, to use the INEE Minimum Standards as a quality criterion for awarding funds and monitoring and evaluating the programs they are supporting.
- Universities, to use the INEE Minimum Standards as a framework for research and academic papers.
- Education practitioners, to use the INEE Minimum Standards as a scaffold of technical knowledge, tools and good practices with which to address capacity gaps and bring actors together at country and global levels.

12 | Establishing Safe Learning Environments

SIMON REICH

War in the twenty-first century preponderantly takes place within states, not between them. Indeed, in the last decade of the twentieth century, intrastate wars outnumbered interstate ones.[1] When intrastate wars occur, huge displacement inexorably follows, mostly in what is characterized as fragile, fractured, or failed states in the Global South. Some displacement leads to self-settlement in new towns and villages, but for millions, such displacement usually results in the creation of Internally Displaced Person (IDP) or refugee camps. Most of the inhabitants of these camps are children, according to one recent estimate:

> Right now, instead of playing games and going to school with their friends, an estimated 20 million children and adolescents are living in makeshift camps far away from their homes, communities and loved ones. They are refugees and IDPs forced by war, natural and manmade disasters, persecution and economic collapse to abandon their childhoods. In the past decade alone, an estimated 60 million children have had to live this way.[2]

This phenomenon is nothing new, but it is a large and contemporary problem. In the first half of 2009 alone, huge numbers of new refugees entered camps in Pakistan and Sri Lanka as a result of intrastate conflicts. The primary focus of this chapter, however, is on the African conflicts that have taken place over the last decade and a half.

Western audiences generally assume that camps provide a safe haven. IDPs and refugees may head for them in the hope that they will be fed and protected upon arrival. Neither, however, can be guaranteed and should not be assumed. With so many children inhabiting most camps, protecting them against violence is a major priority for both the UN and humanitarian organizations. Indeed, the cornerstone of any camp program, part of what UNICEF has termed "child-friendly spaces," has been the creation of a stable, secure environment in recent years. "As a result," write Maureen McClure and Gonzalo Retamal, "refugee camps were increasingly designed around protective spaces for children. Activities and services for the young were located in the safest areas with the most visibility. For example, sometimes they were located near the water supply. Camp traffic ebbed and flowed past children and youth at play. There was safety in numbers. The scene offered chronically depressed adults a visible, tangible symbol for a brighter possible future."[3]

Yet the protection of children from external attack and abduction has not been the highest priority among the international community. While there has been a welcome increase in the discussion around issues such as sexual violence and child protection within camps, relatively limited attention has been given to the issue of camp protection for children from external attack and abduction over the course of the last two decades by either policymakers or academics.

The question of what reduces a camp's vulnerability thus remains largely undetermined. Education within camps stands as a paramount developmental, humanitarian and geostrategic issue. Yet it cannot flourish in the absence of a secure environment. This chapter therefore focuses on the issue of external attacks and child abduction: what has happened and what might be done to address the issue of what makes a camp safe? Providing such safety for children is a prerequisite for the implementation of any educational policy.

IDP and Refugee Camps, Child Protection, and Education: A Study

Although the age of the Internet has greatly expanded the capacity for filing and reviewing of documents on the ground, most data remains anecdotal, and there is a dearth of systematic analysis of both camp security and child abduction. Much of the prevailing literature on child soldiers, for example, focuses on the causes of child soldiering in the broadest sense, is the product of reports by non-governmental organizations, is based on very limited data, and often is careless in terms of the method of data collection.[4]

Between 2004 and 2009, I directed a series of studies that sought to examine the issue of camp security and child abduction. The goal was to gather data on both by building an unprecedented database with the intent of systematically examining the linkage between the two, the dynamics of child abduction and key ingredients that policy-makers need to consider in addressing child safety. The project focused on seven African conflicts.[5] We identified and plotted the geographic location of a total of 1,898 camps in these seven conflicts areas and gathered data for each year that the camp existed, yielding a total of 4,313 "camp years." We were able to record and verify a total of 1,456 attacks on camps. Of the 7,868 documented children abducted, 7,481 (95 percent) were abducted during an attack. Over the course of three phases of the project, we issued two major reports and a series of policy briefs, examining a total of twenty-seven factors that might influence the vulnerability of a camp to child abduction.[6] The results were shared with various stakeholders in the international community including UN officials, various national governments, and interested NGOs. What follows relates some of the substantive findings of this work to the themes addressed in this book.

Nothing Protects a Child as Effectively as an International Peacekeeper

Camps that have a PKO to protect them are far less likely to be attacked than those without one. In one study, where we looked at 1,180 camps that were attacked, 83 percent had no form of protection, while only 17 percent had *some* form of protection.[7] The options for child protection in camps essentially come down to three choices: protection by national governments where the camp is located, international peacekeepers, or forms of self-protection. Some form of screening initiative can be used as an option in conjunction with all three alternatives. Peacekeepers have suffered from much criticism: they are ineffective, corrupt, and lack any sense of commitment. Indeed, stories of their failure in the Democratic Republic of the Congo, Sudan, and Uganda flooded the media in 2008 and 2009. The results from our work are unequivocal, however, in concluding that international peacekeepers provide the best form of child protection.

National governments are, in contrast, the least effective means of protection. Too few troops are generally assigned, often actually encouraging rebel forces to attack the camp so that they can claim a victory, however pyrrhic. Added to this problem is that government troops, entrusted with the protection of civilian populations are not only less inclined to defend refugees of IDPs, but they often attack the very civilian populations they are supposed to protect.

Self-protection seems a liberating and emboldening response. But governments are understandably reluctant to arm civilians (often foreign ones) within their midst as a potentially destabilizing influence. Thus, despite the rich and growing literature on refugee warriors, there is little evidence to support the proposition that self-protection is an effective means to combat external attacks.[8]

Peacekeepers Need to Be Present in Sufficient Numbers to Protect Children in Camps

There are, of course, qualifications to my generalization about the appropriateness of peacekeepers for the purpose of child protection. First and foremost, they have to be stationed in sufficient numbers: a purely "symbolic-sized" force does not offer sufficient deterrence. Long gone are the days (if they ever existed) where a blue UN peacekeeping flag in itself gave belligerents sufficient pause for thought. Indeed, the presence of a nominal number of peacekeepers often sends a signal to belligerents that *there is something worth protecting, so there is something worth attacking at relatively little risk.* The ratio of peacekeepers relative to the number of civilians they have to protect is often nominal and camp populations suffer as a result. Darfur, for example, is often mentioned as being about the same size as France.

Yet in late 2008, the security forces responsible for civilian protection (excluding the military) in France, the Police Nationale and Gendarmes, combined for a total of 224,571. In comparison there were only 12,442 uniformed UNAMID personnel present in Darfur at the end of 2008, just over 5 percent of the French total protecting a population of slightly over six million—in an active war zone. Stated comparatively in a more familiar context, by 2008 each peacekeeper in Sudan was responsible for the protection of 280 civilians; that compared to 88 people per peacekeeper in the DRC.

In this context, children constitute a most desired target in the struggle for resources. Indeed, some earlier research I conducted revealed that access to camps constituted the single most significant factor in explaining the variation in the numbers of child soldiers in these conflicts: the more accessible the camps, the higher the numbers of children abducted.[9] It played a far more important role as a determinant of child soldier rates than other oft cited factors such as poverty levels or orphan rates. Our findings suggested that camps are the epicenters for child soldier recruitment. Education through reintegration programs is vital as a mechanism to break this cycle and ensure that former child soldiers do not relapse into re-recruitment (as discussed later). But protection in sufficient numbers is key to thwarting the creation of that cycle in the first place.

Early Intervention Is Central to Any Strategy

The international community in particular, has long debated the question of early intervention for civilian protection in general, and child protection in particular. Over the course of the last decade, the UN has heavily debated the relevant virtues of early engagement to avert conflict.[10] Our research specifically focused on the question of child abduction, and question of when most abduction takes place.

The findings were clear: most attacks on camps, and child abduction from them, take place in the relatively early phases of a conflict. Although an anathema to the way it characteristically operates, early intervention would provide protection at the most crucial period. Usually, the international community provides camp protection after the worst abuses have taken place.

The data collected about Sierra Leone demonstrates this point for both attacks and abductions. Twenty-eight camps were attacked during the conflict in Sierra Leone between 1996 and 2001. This example is divided into three distinct periods. During the first period, between 1994 and 1996, only two camps were attacked while the tension was slowly rising. In 1997, a military coup occurred after a brief period of relative peace and the signing of several treaties. Shortly thereafter, a violent second period began, exploding from only a few isolated incidents prior to 1997 to eighteen documented attacks in 1998. Between 1997 and 2000, twenty-two of the camps (73 percent) were attacked once per year, while the remaining eight (27 percent) were attacked two or more times per year. Four of those camps were each attacked three times in a year: Koidu (1998), Makeni (1999 and 2000), Kambia (2000), and Waterloo Makeshift Camp (2000). The most violent years for camp attacks in Sierra Leone were, therefore, 1998 and 1999, with eighteen and seventeen documented attacks (67 percent of the total), respectively.

However, in 2001, roughly four years after the attacks had resumed, they began to subside and gradually entered into a quieter third period. Between 1994 and 2000, a total of seventeen different camps in Sierra Leone recorded child abductions. Again from 1994 to 1996, while tensions were rising, there were very few documented attacks. In 1997, when the conflict exploded into full force, so did the number of child abductions as a result of camp attacks. Of the twenty-three documented attacks among the seventeen camps from 1994 to 2001, eight of them (35 percent) occurred in 1998. In 1999, there were seventeen documented camp attacks in Sierra Leone, while there were only four documented attacks in which children were abducted. In total, we were able to confirm 7,839 child abductions. Thus, 77 percent of documented abductions occurred during this period, just two years after the conflict had begun anew. Of the 170 total attacks involving the abduction of children across the entire sample, only 15 percent occurred in Sierra Leone. After 2000, attacks and the number of children abducted declined precipitously in Sierra Leone. The evidence suggests that, had a large, formidable force been employed at the beginning of the conflict to protect endangered camps, it may have helped decrease the number of attacks and children abducted.[11]

Mandates Are Key for Success

International peacekeepers have to be given an appropriate mandate. UN Peacekeeping missions have formal mandates that legally define their duties and obligations. Peacekeeping mandates fall under either Chapter VI or Chapter VII of the UN Charter. A Chapter VI UN mandate is associated with traditional methods of resolving conflict without the use of force, such as mediation, negotiation, conciliation, and fact-finding. It does not expressly call for the physical protection of civilians. In contrast, a Chapter VII mandate authorizes the use of coercive measures including, but not limited to, sanctions, embargos, and, if necessary, military intervention. A Chapter VII mandate may specifically authorize the use of force to protect civilians. While a mandate is intended to clearly define the role of the mission, its implementation is inevitably dependent upon the interpretation and enforcement of each force commander as well as the context of each conflict.

Simply stated, too often mandates fail to incorporate key components regarding the use of force, rules of engagement, child protection, and the role of UN forces in the protection of borders (to safeguard refugee camps along them). Of the eleven UN mandates we examined, only four had any component regarding child protection.

By way of comparison, the mandates for UNAMID (Darfur), MONUC (DRC), and ONUB (Burundi) contain language that specifically tasks these peacekeeping operations with maintaining and/or reestablishing border security. Our analysis undertook a detailed analysis of conditions in these three states in an effort to better understand the relationship between peacekeeping operations, border security, and the protection of camps. The ONUB mandate contains a number of provisions that explicitly pertain to border security. Specifically, it compelled the peacekeeping mission to monitor and prevent the free movement of small arms across the Burundi/DRC border. Further, the Security Council required ONUB to both protect refugees as they crossed the border and to prevent armed combatants from crossing the border.

The Secretary-General was also asked to conclude agreements with neighboring states allowing ONUB forces to cross into their territories when in pursuit of militants. Though not evident in any other mandate examined, this provision is especially important for the protection of civilian populations. An important commonality between the MONUC and ONUB mandates is that each specifically calls upon the two peace-keeping operations to coordinate their efforts along the common Burundian/DRC border. However, neither mandate includes a call for diplomatic efforts to attain transit authority for the pursuit of militants across the border. Although the Burundi region is still considered unstable, the ONUB mission, unlike the MONUC mission in the DRC, entered the conflict, helped restore peace, and left within a relatively short period of time. How does border security—or the lack thereof—translate in terms of violence? Our research revealed that slightly less than 4 percent of all attacks on civilians in the region occurred on the Chadian side of the border in 2004. By 2006 that figure had increased to 19 percent, and by 2008, the percentage of attacks occurring in Chad grown to 33.33 percent.

When Is "Enough" Really Enough?

Related to both mandates and border protection is the question of the appropriate resourcing of peacekeepers. There are three related problems. First, the specifications of a mandate are often inconsistent with the challenges that a peacekeeping force faces on the ground in the protection of children. Second, peacekeepers often suffer from a lack of munitions, equipment, and logistical support to fulfill the requirements of missions. Third, they often have insufficient numbers and training of personnel. As Ramesh Thakur and David Malone have argued, "clear mandates and goals, matching military and financial resources"[12] are required. Mission requirements, critics claim, are often "undersold" to ensure Security Council support and overcome their reluctance to intervene. A July 2008 report by the Darfur Consortium, for example, found that underfunding has had a severe impact on the UNAMID mission in Darfur. It states, "The force lacks critical resources, leaving the people of Darfur, humanitarian agencies and even its own peacekeepers vulnerable to ongoing attacks and extreme violence."[13]

These criticisms have become particularly acute given the upsurge in violence in the DRC and the inability of the UN mission, MONUC, to offer adequate protection. In its November 2008 report, the Office of the Special Representative of the Secretary-General for Children and Armed Conflict states, "the outbreak of renewed fighting in North Kivu and parts of Oriental province, notably in [the] Ituri district and Dungu territory, in September and October 2008 has led to a resurgence of incidents of violations against children. This is posing serious challenges for the implementation of the reform of the security sector and the comprehensive strategy for stability and security."[14]

In 2000, given the UN's failures in Bosnia and Rwanda, then–UN Secretary-General Kofi Annan commissioned a high-level group of experts to assess the UN system's shortcomings and provide candid, pragmatic recommendations for change. The Report of the Panel on United Nations Peace Operations, commonly called the Brahimi Report, offered an in-depth critique of the conduct of UN operations. The report asserted that vague mandates coupled with overly optimistic assessments of the situation on the ground led to inadequate deployments with unclear objectives. Furthermore, mandates too often exceeded the resources available, creating an operational gap between expectations and actual mission capabilities. The report urged the UN Secretariat to lay out both the resources and procedures for missions operating in a conflict zone, including the importance of reporting potential gaps between mandates and capacity to the Security Council. The panel further stressed that many complex operations require the will to use force if necessary: "Rules of engagement should be sufficiently robust and not force United Nations contingents to cede the initiative to their attackers."[15]

Peacekeepers and DDR Programs:
Where Protection and Education Meet

The linkage between mandates, resources, child protection, and education converges in two clear places. The first is camp security and the creation of child-friendly spaces as described earlier. The second is with disarmament, demobilization, and reintegration programs (DDR) where the education of demobilized youth plays a central role in their rehabilitation and integration process, serving as a bulwark against re-recruitment.

Most failed and fragile states enter an interminable cycle of intergenerational violence. Paul Collier has characterized this phenomenon as the "conflict trap"—political instability, violence against civilians, and a massive retardation in economic development.[16] The struggle for resources generates an unending dynamic loop—there is insufficient time for prosperity, stability, and (preferably) democracy to establish itself in a society before it slips back into war. Education is a key component in breaking this cycle: forming an intrinsic part of the peacebuilding process.

Violent child soldiers, unless sufficiently educated when they demobilize, easily slip into criminality or re-recruitment. In addition to inducing psychological trauma, a violent childhood reduces healthy educational opportunities, leaving some form of militancy (in a gang or rebel group) as the only viable career path in later years. Violence thus becomes a way of life. Reputedly, the Taliban's current leadership initially learned to fight against the Soviets in the 1980s. Reflecting this cycle, in Afghanistan, a fourteen-year-old was responsible for the first killing of a NATO soldier—just one of the estimated eight thousand child soldiers who do or have worked as part of the Taliban's forces in the last few years. Such children often become suicide bombers, as the case of the LTTE in Sri Lanka illustrates, or susceptible to terrorist recruitment.

Educational programs therefore play a key role in breaking this cycle. *Education as a policy instrument serves a humanitarian, developmental and geo-strategic purpose.* Unfortunately, most national governmental policymakers ignore the importance of the last component. If they did not, perhaps they would give it more attention.

In administering educational programs, our findings suggested that UN forces are indispensable in terms of the process of reintegration of youth (especially former child soldiers) through educational and in the aftermath of war. Yet child protection and education were too often omitted as part of the mandate itself, leading to confusion on the ground regarding their role.

Of the seven current UN peacekeeping missions in Africa, five include elements of DDR programming in their mandates. These programs are often regarded as playing a potentially important strategic role in consolidating an end to violence and enhancing the rule of law. While the role of peacekeepers in the evaluation, monitoring, and implementation of DDR programs may differ according to the conflict, DDR programs can be a vital link in strategic peacekeeping— between peacemaking and peacebuilding—by facilitating an immediate reduction in the level of violence and facilitating a long-term return to peace and prosperity through education as a tool of reintegration.

The case of Sierra Leone aptly demonstrates both the potential significance of DDR programs and the role that peacekeepers can contribute to such programs. The UNAMSIL mission in Sierra Leone, commencing in 1999, included a mandate that provided for the oversight and implementation of a Joint Operation Plan to implement a DDR program involving several agencies including the Government of Sierra Leone, UNAMSIL, UNICEF, and the Economic Community of West African States (ECOMOG). An educational component was a central pillar. When violence resumed in May 2000, the DDR process was halted for several months until combined efforts later that year led to its reinstitution. This last phase of the DDR program, beginning in May 2001, was characterized by an increased commitment to the peace process by all armed groups. Over the ensuing three-year period, the program proved effective in contributing to the curbing of violence. UNAMSIL's role included providing security, establishing reception centers and weapons storage centers, implementing screening and registration and providing resources for relocation and family reunification. The program eventually disarmed and demobilized 71,043 combatants. Sierra Leone correspondingly experienced a rapid decline in attacks on civilians. International forces were key to the mission's success. They do not always work. But DDR programs can have a complementary role in the peacebuilding process and should therefore be implemented by peacekeeping operations when possible and pertinent.

Conclusion: Youth Bulges, Violence, and Education as a Security Issue

Policymakers usually see education as a humanitarian issue and often as a developmental one.[17] Sadly, even countries such as Liberia that have come out of a intrastate war relatively peaceably suffer from high unemployment rates—in Liberia's case, over 80 percent. With the youth bulge symptomatic of many countries, they occasionally view education as a demographic issue. They rarely view education as a geostrategic issue. Yet education programs have enormous strategic value in both crisis situations and the postwar construction process. They increasingly form a bulwark against both recruitment and recidivism to gangs and belligerents in failed and fragile states. Protection and education are thus intrinsically linked.

Most recruitment by armed belligerents in African conflicts takes place from camps, puncturing efforts at the construction of child-friendly spaces. These programs therefore do not take place in a vacuum. They do so in conflict zones—and the record suggests that, despite all their flaws, international peacekeepers are the most capable of protecting children and creating or sustaining child-friendly zones.

Of course, PKOs fail when mandates are vague or too limited, when missions are underresourced, and when troops are insufficient in terms of both numbers and training. They work, however, when the international community commitment matches a child's needs.

13 | Psychosocial Issues in Education

ARANCHA GARCÍA DEL SOTO

"Psychosocial" work tends to be labeled "soft" when compared to other approaches in humanitarian aid. Different programs on sanitation, shelter, and nutrition address material and physical needs that have a direct and clearly measurable impact on the lives of internally displaced persons (IDPs) and refugees. It also tends to be commonly accepted that education programs are important in emergency contexts, although their overall impact is longer term oriented. Psychosocial work has been defended recently, despite the absence of hard evidence, because its issues focus on crucial considerations like "local empowerment and the restoration of dignity," as well as the consideration of the specific circumstances of each context, culture, and its peoples.[1] Participation and cultural sensitivity are global aspects that emergency work acknowledges, despite the risk of using them just as "token words."

The populations of IDPs and refugees worldwide have experienced and continue to experience physical hardships, trauma, and pain, but along with all of these we find resiliency. It is the resiliency that ultimately allows the so-called beneficiaries/survivors or right-holders to overcome the tragedy and continue to live their lives in the best possible way. Resiliency is closely tied to the consideration of the mutual support and interaction between individual and community wellness. It permeates every single emergency program.

It is widely acknowledged that education in emergency situations provides the "physical, psychosocial and cognitive protection that can be both life sustaining and live saving."[2] Collectively, education pro-grams provide safe spaces to ensure the gift of learning and the hope it brings to those communities that have experienced warfare or natu-ral disasters. Individually, it allows for the identification of persons who have been more severely affected and might experience a slower recovery because of the traumatic experiences undergone (or those who had previous learning or mental problems that may become more exacerbated due to the crisis, thus causing one to fall behind even deeper after it).

"Education mitigates the psychosocial impact of conflict and disas-ters by giving a sense of normalcy, stability, structure and hope for the future," remarks the Inter-Agency Network for Education in Emer-gencies (INEE).[3] The wide variety of education programs touches not only on the psychosocial experiences of the population affected, but also plays a major role in advocacy and sensitization, providing "the knowledge and the skills to survive in a crisis through the dissemina-tion of live-saving information, such as hygiene, landmine safety and HIV/AIDS prevention."[4]

From a long-term perspective, few programs seem to have a more solid impact than the educational ones. They build "social capital, enhance social cohesion and support conflict resolution and peace building . . . contributing directly to the social, economic and political stability of societies."[5] This point is so salient that the violent actors, interested in the continuation of violence and the destabilization of social peace, attack education by targeting teachers, students, and in-stitutions that implement education programs.[6] Destroying education destroys the hope and future of the enemy, which can prove to be a highly effective weapon of war. Disrupting children's developmental progress breaks down not only the individual child and his family but also, as a collective phenomenon, interrupts the progress of the very fabric of society.

Hope and a minimum sense of control over people's daily lives form the basis to promote psychosocial well-being. I personally learned that the school system the Slovene government set up for Bos-nian refugees who took shelter in the country in 1993 was a major source for normality and recovery. It was not exclusively the formal education system for children, but also many other informal education activities that volunteers and organizations were implementing in refugee settings for people of all different ages. These cases proved for me the structural connection between formal and informal educational activities and psychosocial well-being. Giving survivors different options to get together to talk and exchange experiences provides a safe social environment, which reinforces a sense of identity (both individual and social: "What is left of the person I used to be? Why did this happen to us?") that is usually attacked during violent times.

Inger Agger worked on the evaluation of the psychosocial programs implemented by the European Commission Humanitarian Office (ECHO) in Bosnia and Herzegovina, revealing that the most valuable services in the eyes of the beneficiaries were the psychosocial programs that facilitated group activities.[7] One of the paths to recovery is being given an opportunity to get together with other members of the community in a place that feels safe enough, to share experiences. Being both physically and psychologically accompanied proved to be very important in promoting well-being. This "simple" need, which stems from the basic human need to share and belong, is one of the crucial aspects of the recovery process, both at the individual and the collective level. It was also found that fulfillment of this need facilitated the creation of groups by these victims that worked in favor of their own goals, as well as for other beneficiaries who have gone through similar experiences.

Education and psychosocial approaches are intertwined, in the following goals:

- The hope connection: Education programs are highly valued by survivors because they channel their hopes for a better future and provide better options for future generations. Hope is an essential value in well-being, the core desirable impact of psychosocial programming. Having goals allows one to be more present and responsible each day, while facing the possibilities o a future.
- Overcoming frustration: It is also the case that learning and overcoming learning difficulties are essential for the well-being of survivors. Ideally, different types of learning should be adjusted to individuals with diverse learning styles and cognitive abilities,

different ages, and different backgrounds (e.g., rural/urban, mental health problems, or different socioeconomic status). There is no learning if the person does not have a minimal mental and physical stability and does not feel safe. Conversely, there is no real wellness if the person feels stuck and not exposed to new learning situations.

- Sustainability, empowerment, and ownership: Formal and informal education involving survivors (through participation and tools like 'training of trainers' workshops, which can be replicated after the expatriates leave) certainly ensures a much better sense of ownership. And ownership by the local actors and institutions is the strongest warrant toward sustainability, maintenance, and progress of programs that prove positive after the emergency is over.
- Prevention: School is a psychological protective factor for children and learning is a protective factor for all human beings. Curiosity provides people with the energy and connections that expand their horizons and help set new goals, while developing new interests on top of the already existing ones.
- A common endeavor facing the future: Formal and informal education takes place around a network of actors; students, trainers, teachers, parents, community leaders, community members, bystanders, and so on. All of them play major roles in the reconstruction of the social fabric. It is through their interactions and the exchange of best practices that we can aim for a better world in the long term.

In this chapter, we address some recommendations and specific examples of education programs in emergency contexts, analyzing their psychosocial aspects and well-being goals, while focusing on the actors involved (parents, teachers, students, and trainees with different needs), and formal and informal educational activities. We give special consideration to the groups of children and adults, the subjects of education, that have mental health problems and/or learning difficulties, both before and after (as a consequence of) the emergency situation, due to warfare or natural disasters, to finally comment on the need to consider psychosocial and protection aspects transversally, in all emergency programs.

Approaches: Psychosocial Lenses and Approaches when Creating Education Programs

Within the realm of psychosocial work, we can find three major approaches: the individual approach (with more focus on mental disorders), the community approach, and the human rights approach. They are not mutually exclusive and have many overlapping qualities. In the education programs we tend to initially frame our psychosocial goals with more of a community approach; after all, education tends to be implemented in groups, and collective approaches toward learning make sense in the global recovery post-emergency contexts. Ideally, we should also be attentive to those cases that require more individual care: children traumatized by the violent experiences they have gone through during the emergency situation, or people suffering from mental illness prior to the emergency (who are at risk of getting worse or falling out of the collective efforts in the post-emergency phase). From a human rights approach, education is a right linked to social development, communication and intergenerational equity, and for this reason should play a major role in promoting peace, given the circumstances of the specific contexts, and human rights.

In the Inter-Agency Standing Committee (IASC) guidelines on mental health and psychosocial support in emergency settings, four common functions are emphasized:

- Facilitating conditions for community mobilization, ownership, and control of emergency response in all sectors.
- Facilitating community self help and social support.
- Facilitating conditions for appropriate communal cultural, spiritual, and religious healing practices.
- Facilitating support for young children (0–8 years) and their caregivers.

The last two are directly connected to education.[8]

The contents of education should be, in all circumstances, transpired by human values: attitudes and behaviors that promote peace, tolerance for diversity, and mutual support. An adequate peer culture can have an enormous positive effect in all children and adolescents. It helps to provide those involved with a feeling of safety, connected to a sense of the cultural and social identity that is so prevalent at these stages of live. A good initial school system for children provides a solid foundation for their well-being and further development, and better equips them with a sense of belonging and the joy of learning while growing up.

Education in emergency situations, during warfare or in the early aftermath of war or natural disasters, has unique features connected to the four functions assigned by the IASC guidelines. When initially programming and organizing the educational set up it seems advisable to first bring people together (parents, teachers, a good representation of community members) and then agree on the organization and set up of the education programs. It is important to continue the dissemination and sensitization work surrounding these activities (talking and sharing, anticipating pros and cons, and distributing responsibilities) with the rest of the community, to ensure the community has a sense of real involvement and ownership. Education programs should go hand in hand in emergencies with the logistics, infrastructure reconstruction, shelter, and nutrition programs. All of them have an impact in the social support dynamics of the communities affected and are directly connected to the sense of control on their lives during these crucial changing times. In a true community approach, the formal school system for kids needs to be negotiated among the community members. Ideally, they should decide on the informal education activities, assigning clear responsibilities to monitor these processes, making sure that progress is discussed as the emergency evolves, therefore enabling them to change and improve things, as the process moves along. Community members should also make sure that those with trauma and learning problems are given special care.

The first version of the Sphere Project did not include educational or psychosocial guidelines, but a later revised version (2004–2005) did emphasize the importance of maintaining religious and cultural events along with the importance of schooling: "As soon as resources permit, children and adolescents should access formal or informal schooling and normal recreational activities." These guidelines also underline the importance of providing medication and intervention to those psychiatric cases that need it, and to consult the community regarding decisions on where to locate sanitation facilities, religious places and schools. In the debates preceding these guidelines, psychosocial workers with experience in emergencies in different settings shared their experiences regarding schooling and setting up informal educational activities and the psychosocial impact in the community. All agreed that it is not only the children that benefit from having a place and hours filled with educational activities, but also their parents, who are able to enjoy more private, quiet or working time, thus spreading to the community overall as the dynamism and optimism are contagious. A repeated argument heard within refugee and internally displaced communities is, "Our kids are/are not able to go to school," followed by comments on how this basic fact labels their overall situation as more or less hopeful.

Within the community mobilization strategies to set up education programs, the role of the community leaders (accountable in the eyes of most community members and also holding good communication with the external agencies) and the selection of teachers are crucial. In many circumstances, people with no substantive experience on teaching can undertake responsibilities in the education programs agreed, as long as their responsibilities and resources are clearly assigned and they count on the support of other actors involved in the humanitarian efforts.

The Psychosocial Working Group (PSWG) states in its Conceptual Framework that three core domains define the well-being of individuals: human capacity, social ecology, and culture and values.[9] Human capacity refers to the knowledge, skills, and physical and mental health of the person. "In these terms, improving physical and mental health, or education and training in support of increased knowledge, enhances human capacity, and thus psychosocial well-being." It also describes how there is evidence linking "mental health outcomes to the presence of effective social engagement, but wider cultural and programmatic concerns also justify the specification of social ecology as a discrete domain underpinning psychosocial well-being." The third domain, culture and values, "is recognized in its own right as a third key determinant."[10] These three core domains are intrinsically connected to education. If only all emergencies had the economic, physical, and environmental resources to set up formal and informal education programs effectively, the recovery phase would be faster and sustainability in the path toward a better future would be more feasible.

Children with Problems

Ideally, within all these community mobilization strategies in emergency situations, there should also be enough human and material resources to provide clinical treatment to those with "mental health problems" in the impacted population. Previous psychological problems, or "learning difficulties," associated with cognitive and emotional issues resulting from the traumatic experiences (during and after the emergency situation) also pose an enormous burden in the individuals suffering from them. Formal education is secondary to children who have gone through traumatic experiences.

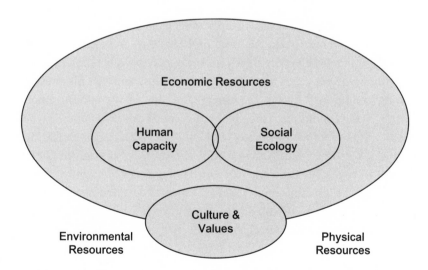

Figure 13.1. Conceptual Framework of the Psychosocial Working Group, 2004.

In many cases, when possible, it would be more favorable first to group kids with similar problems, before aiming to integrate them. It is recommended to deal with trauma before integrating them in a more formal education process. Guilt and shame are the two most powerful feelings impregnating the lives of these individuals, and fear follows those two. We need to acknowledge and work through those impeding feelings (shame, guilt, extreme sadness, and fear), making sure that the children with problems feel less threatened, safer, and a bit more willing to be integrated with other learning groups.

In many cultures, there are not Western diagnoses for mental health problems as we have them in the West, but the signs of trauma and pain can be easily detected by community members: nightmares, sadness, memory problems, impaired concentration, aggressiveness, loss of interest, inactivity, apathy and numbness, mistrust, various psychosomatic complaints (loss of appetite, stomachache, headache), and regression (going back to early stages of development and showing more dependency, earlier speaking habits, bed-wetting), and so on. If these symptoms are of a reactive nature to the "craziness" of the emergency situation, they "will pass in one to three months. But many children have been afflicted more deeply by tragic losses and extremely painful experiences."[11] For many of them, having lost a parent or having witnessed deeply cruel and shocking scenes, for these "emotionally wounded," the probability of these disorders appearing in their behaviors depends on the survivor's personality, pre-emergency experiences and social relationships, and "on his present living conditions" during the emergency. Within the current emergency living conditions, "the most important factor is his family: who the child lives with . . . who was left behind . . . what the psychological state of the other family members is, the relations inside the family and how much attention and support he is given by the family."[12]

Many of these children should be expected to show new learning difficulties, perhaps on top of previous ones. How can the school, the teachers, the parents, and the community help the children, overall, without leaving behind those with the most acute learning problems and trauma? In many emergency situations there are not enough teachers or psychosocial professionals available (only recently, in the context of the post-tsunami Sri Lanka, there were fewer than a few dozen psychologists for the thousands of citizens affected, and this figure includes the Sri Lankan professionals returning from overseas to help). When human resources are scarce, the role of the grassroots community members (mothers, elderly people, union leaders, and so on), volunteers (locals and foreigners) and teachers is definite. They can be trained and conduct trainings themselves (in a rolling/multiplying/"snowball effect" way), to educate the community to detect and acknowledge the emotional and learning problems of the children and how to better address them. Anica Kos describes some of these workshops and the guidelines given to teachers and family members for the thousands of refugees from Bosnia who were sheltered in Slovenia in the Balkans conflict during the 1990s.[13] Naomi Richman, based on her experience working with children in Mozambique, trained teachers from Bosnia and Herzegovina and recommends that in order to help children speak about their experience, parents and teachers should plan enough time for the conversation. Teachers and parents, according to Richman, would ideally find a delicate balance of creating a space to speak about the child's feelings, without creating pressure to do so, while always showing the child respect and encouragement.[14]

Regarding the programming, it is recommended to create activities that promote children's confidence, by organizing other activities apart from the formal learning ones, always avoiding criticisms and always encouraging them. "All children need the possibility to express themselves though drawings, poems, singing, texts and discussions. Some children will take on the responsibility for organizing school and out-of-school activities. This is the best remedy for the feeling of helplessness, which is so very characteristic of, and damaging, to refugees."[15] To accept, encourage participation, and accompany these children in their search to feel better is a key for success, not only in refugee or emergency settings, but also in most situations. In the next section, we describe some activities linked to artistic expressions, but it is commonly accepted that art group sessions for kids with problems will be very beneficial to facilitate their emotional expression and promote their integration with others kids their age. Painting, singing, storytelling, or music sessions can be extremely beneficial for the kids and their caregivers.[16]

Holding a holistic approach that views education in terms of human values is key when assuring the protection of the most vulnerable. We should reinforce the positive approaches that families and communities already master to protect their members suffering from mental health problems, and to design new protective strategies where there are none, promoting the education and sensitization toward individuals with mental health problems.

The well-being of the families and daily caretakers, the ones really supporting the population with mental health problems tends to be forgotten, when in most of the cases, they are the real "experts" and ones who should always be consulted to set in place the appropriate programs. Focusing on their well-being in a "care of helpers" approach is also highly recommended, whenever possible. Social emergencies can be great opportunity windows for advocacy: promoting and benefiting from the experiences and the visibility of the work and efforts of the teachers and families of children with mental health and learning problems.

Formal and Informal Education Activities, and the Voices of the Actors Involved

The creation of a formal school system will certainly provide a sense of normalization in the context of the emergency situation. Keeping in mind all different ages (kindergarten, primary, and secondary school, whenever possible), the formal learning courses of a schooling setting will stimulate the mental and social progress of children and adolescents and will benefit the community as a whole. In emergencies, it becomes necessary to train teachers to better detect emotional and learning problems in order to treat those children, following some of the guidelines mentioned in the previous section. It is also very important to consider the well-being of the teachers, who sometimes are survivors themselves, and usually feel the additional pressure and hardships of the losses and the uncertainty of the emergency context.

Parents should also be kept inside the loop, playing important roles in the decision-making process and in the monitoring of the progress of the kids at school. The teachers should always be there for the parents, and the parents should work as closely as possible with the teachers. Mobile support teams can be a very useful strategy to respond to the lack of professionals and the dispersion of the affected emergency populations. These teams can be composed of counselors, physicians, and any other professionals and volunteers who can convey and exchange information on needs and support strategies, among the populations affected and other resource teams in different geographical areas.

Along with the formal educational activities, the informal activities cannot be underscored enough, since they seem to be essential for all kids; those with learning disabilities and those who can also easily attend the regular classes. Both formal and informal activities are crucial for child development and the reinforcement of a peer-group sense, despite all the emphasis we tend to place exclusively on formal schooling.

Informal activities are particularly important for children with emotional and learning difficulties, as these activities may be the only way such a child can participate as part of a group. Artistic and recreational activities provide the opportunity of being liked (and generally, we all like to be liked). Through the expressive production of the arts, we work through emotions and feelings that are basic components of personal growth, adding to the more reason-based learning (in some countries still basically memorization). Many times, the products and results of these artistic, sportive or other informal activities allow the person to express herself and to show her understanding of the situation, or to feel better integrated into a group. It can also be a way of expressing guilt and shame, and this might allow for new insights into how people feel about their experiences. This possibility is very important since managing that guilt is a decisive step toward well-being. People might be blaming themselves for things that are out of their control, and thus reinforcing the message to them—"It is not your fault"—might help the healing process.

Some of the informal activities that can be conducted in refugee, internally displaced, or other emergency settings are: gardening, hairdressing, fitness-related activities (swimming, aerobics, yoga, or other popular sports like soccer, as long as we don't replicate group tensions and we avoid making them too competitive), massage and basic physical care for babies and children (encouraging the participation of other family members), book clubs, and music and dancing for teenagers. With these age groups, also graffiti and "photo-voice techniques" (having them take pictures to later comment or work around those pictures) are also very popular. Other activities like theater, dance, and workshops on some of the pressing needs (such as hygiene, sexuality, addictions, and so forth) can work well when helping the person feel more active, busy, and sociable, when given the opportunity to share experiences and support others. Activities designed to put them in touch with nature, such as walks, excursions, learning about the outdoors, and the like, can also be calming and extremely gratifying.

All of these activities should of course be adapted to the specific cultural norms of each context, for in some scenarios it will not be appropriate to mix genders or to speak up on certain topics. Also, when the illiterate population is abundant, teachers will have to come up with ways to implement activities that everybody understands (graphics, drawings, and so on), which these people can follow. Overall, the mixing of different generations in some of these activities (like music workshops or storytelling) tends to work well, and it encourages new understandings between the elderly and youth, providing all with a sense of worth.

For the informal activities more connected to the problems and realities faced by the community during the emergency situation (for example, workshops on material resources needed or human rights advocacy), it is recommended to count on community leaders (elderly, tribal leaders, teachers, or respected and well-accepted foreigners). They can monitor the internal dynamics of the group and should play a conflict-mediator role, bringing together different proposals and suggesting new community programs when answering to the emerging needs or concerns that the community members express in these workshops. In Angola, a local psychologist uses the term "*facipulaçao*" ("facipulating") to define the two main functions of these workshop's mediators: on one hand, they facilitate exchange and interaction, and on the other they manipulate "a bit" to extract the most useful suggestions and contributions, for the community to be able to agree and set up further goals that most of them can benefit from.

The main actors involved in the educational activities during emergencies are the different children (with their different needs and learning styles), the parents, the teachers, the volunteers promoting informal activities, the rest of the community members, and the local and external agencies in charge of the institutionalization of the education programs. We have seen how each group has different and similar needs, and how the interactions and communication is crucial for the adequate progress of the education activities, always aiming toward the well-being of the children. It is important to make sure that parents and teachers communicate well, and to try to find the time and appropriate spaces for them to share their needs and concerns.

In the next paragraphs I attempt to convey the voices of some of these actors through selected testimonies. There is no one better than the individual to talk about the effect of education activities on his or her feelings and psyche and to express the relevance of the relationship between the two.

This is the testimony of a Bosnian teacher, a refugee himself, who worked in Slovenia during the first years of the war:

> Besides all those things, we did much more for the refugees because the school was the only institution in Slovenia that they considered as their own. That was the reason why they came to school with all the questions they might have had. We have done a lot of things also with the humanitarian aid workers, as in connecting refugees with their relatives who stayed back in Bosnia-Herzegovina. . . . We were doing everything that we could and that we knew might provide some help to people. The principal role of the teaching staff was: The parents who come to us for help, but specially if it concerns a child, or if a child comes himself, must never go

away with a feeling that there is nobody to help him, that there is nothing that can be done for him and that he is of no importance to anybody.[17]

Meho was an eleven-year-old student, a Bosnian refugee in Slovenia, who left for Germany, and the following is one of the letters that he wrote to his teacher, Adveta:

Hello teacher Adveta. How are you and my old school friends? You will start the next school year without me. I'm sorry to have left you. We came here to Berlin a few days ago. In the beginning I was so bored here, but now I'm not. But I still think of you and Trbovlje all the time. I would like to be with you at least for an hour. I thought I'd stay with you, but destiny didn't want me to. Maybe we'll never see each other again, but we'll keep on sending letters. I'll be going to the German school now. But I won't spend as much time with the teachers as I did with you. . . . I will always remember the excursion. . . . If you move from your apartment, send me your address immediately. We haven't got the apartment yet, so I can't send you my address. I wanted so much to celebrate Christmas with you, but I couldn't. . . . I haven't gotten to know a lot of people here until now, I only know the ones I've had to meet. Here in Berlin all the land is flat. The city is still being built. There will be Olympic games. Berlin is such a big city, that you can't orientate. I would like so much if we could take a walk together. Just like when we went jogging together. You can't compare Trbovlje to Berlin. But I will love Trbovlje for the rest of my life anyways. And so the boring days pass by. I don't know what else to write. Bye, Meho.[18]

As a last example, in 2005 I personally got a call from a Slovenian friend, a former refugee center director, telling me that a mother and her daughter whom I met as Bosnian refugees back in 1993 had been living in the United States for more than twelve years, and that the mother wanted me to call her. I remembered them well; the mother was an energetic and lively woman who had suffered sexual abuses while in detention before arriving at the center. The daughter was suffering from nightmares and regressive behaviors. We talked. They were well, doing fine. The most impressive thing for me was to find out that the daughter was starting medical school, and that she hardly could remember anything about me and my work as a volunteer at the center in 1993 and 1994 but the "cabarets" and dancing parties and the swimming activities. I was expecting to hear something more, since I had tried to have a different impact through my psychology training and other more formal activities with the teachers at the center. Yet, I felt happy to have been part of the fun that a thirteen-year-old, going through a very difficult time, in the midst of the sad and uncertain situation that she, her mother, and other neighbors and peers from Bosnia were going through those days. She has a very different, and much better, reality now.

Calling for a Holistic Approach in Emergencies That Integrates Education and Psychosocial Work

This chapter tries to present practice-informed theory regarding the connections between formal and informal education programs and the psychosocial well-being of survivors in emergency situations. The PSWG defines "well-being" as the connections between the psychological aspects of our experience (thoughts, emotions, and behaviors) and the wider social context (relationships, traditions, and culture): "These two aspects are so closely intertwined in complex emergencies that the concept of 'psychosocial well-being' is probably more useful for humanitarian agencies than narrower concepts such as mental health."[19] And ideally, we should be able to aim for both when setting up education programs in emergencies; a more collective, community-oriented approach that can promote the well-being of the largest possible number of survivors (for example, a formal schooling system), along with the detection and treatment of those specific cases with learning problems (such as artistic informal activities) or those that need professional mental health support (for example, psychiatric treatments including medication, individual and group therapy talk sessions, as well as other integrative, informal collective activities).

In reality, specific educational activities and attention programs for children with learning problems or mental health problems are rarely found in emergency situations. This is either due to a lack of material and human resources, or because there is a lack of consideration in the cultural context of the emergencies of these psychiatric problems. However, the suffering of the adults and children affected by disorders such as attention deficit disorder (ADD) or bipolar disorder is obvious to everybody around, and it may be mitigated with some professional interventions adapted to the family, community, and cultural circumstances.

Through education programs, children and people have the opportunities to escape an environment full of the tragic memories and sad stories of what has happened to family and community members. When educated, "a child's thoughts are engaged in the pursuit of other topics. His intellect and experiential world gets broader and fills with new ideas. The school broadens his horizons and thoughts,"[20] which otherwise might just be confined to the obsessive memories of the traumatic experiences connected to the emergency. Education programs provide survivors with a sense of control over their daily lives; and they normalize, stabilize, and recognize small daily efforts, while encouraging discipline and working toward more long-term oriented goals. They can also become replications of social contexts in which things are more like they are ideally supposed to be: safer, happier, and more gratifying. In this context, the values of peace and tolerance can be promoted and taught in a conflict prevention manner.

Overall, education is one of the most valued opportunities, especially for the people that have lost most of their material belongings: "They couldn't take away what I know, and what's in my mind," was the way many survivors of the conflict in the Balkans described the different impacts the violence had on themselves. This sense of ownership, in addition to the hope provided by expecting better opportunities to apply this knowledge and abilities, is truly empowering for the victims of violence.

Through many of the education activities people experience that they are not alone. It is not only the mutual support within a group of individuals that have experienced similar losses, engaged in similar learning goals, but also the accompaniment of the teacher and the people managing the education programs and the informal activities (parents, volunteers, locals and foreigners, external actors within the international agencies, and so forth). This "get together factor" might seem minor, when compared to all the needs experienced by the survivors, but it is not.

Also, many of us as humanitarian aid workers can play the role of the witness, as external agents, by disseminating these realities in order to prevent them in the future. Emergency settings can become small worlds of hope, where disciplines work together to bridge theory and application; where external and local actors exchange in a participatory manner; where gender and generational differences are managed in a rights-based approach, combined with a culturally respectful way of doing things. We should constantly aim to improve the situation of the survivors and affected populations, as part of the effort to create a better and more humane world.

In World War I, 90 percent of those killed or injured were soldiers. Today, in the ever-increasing armed conflicts and natural disasters that cause such havoc around the world, 90 percent of the victims are civilians. The most vulnerable, women and children, are subjected to purposeful, terrible harm. To appreciate the scope of this human calamity, and to understand the indispensable role that education plays in healing the wounds of war, and preventing the spread of this evil epidemic, I have invited humanitarian workers from some of the worst afflicted areas of the world to share their personal stories and, even more importantly, their suggestions for improving the way we serve children.

14 Education in the IDP Camps of Eastern Chad

GONZALO SÁNCHEZ-TERÁN

Since 2003 eastern Chad has been the stage of one of the worst humanitarian crises of our time. The outbreak of the Darfur conflict led to a massive and ongoing movement of people toward safer areas. Six years later, more than 250,000 Sudanese refugees still live in twelve refugee camps along the border. The rapid response of the humanitarian agencies provided refugees with a decent level of services considering the harsh living conditions in the Sahel region and the insecurity that has prevailed since the proxy war between Sudan and Chad started. The World Food Program (WFP) has continued to deliver food to the refugees despite logistical constraints and the United Nations High Commissioner for Refugees (UNHCR) together with other UN agencies and humanitarian NGOs have taken care of other assistance needs. Even primary education, a sector historically neglected during humanitarian emergencies, was properly fulfilled. Media attention and the funding that it mobilized were instrumental in the creation of livable refugee camps.

In 2006 armed conflict erupted between the Chadian Government and internal rebel groups. This, combined with the ongoing conflict in Sudan, only worsened the tribal rivalries in eastern Chad, causing widespread violence and the displacement of thousands of Chadian peasants. The international media unfortunately was looking elsewhere.

IDP camps sprang up as villages were destroyed, people were killed and families were forced to flee. By November 2008 the number of IDPs in eastern Chad had risen to more then 175,000. The initial response to this new crisis was slow and insufficient. Massacres in Chad were attributed to a spillover of the Darfur war and Chadian IDPs were considered a collateral damage of the Sudanese conflict. There seemed to be no need for further analysis and little need for further aid.

The government of Chad was initially unwilling to acknowledge the gravity of the humanitarian crisis. Doing so would have meant recognizing the existence of a serious internal conflict in the country. For the international community, the growing civil war in Chad was embarrassing, for France and the United States supported President Idriss Déby Itno while gross violations of human rights were being committed by both sides with no hint of democratization of the regime. As a result, a year and a half after the opening of the first IDP camps in eastern Chad, few international NGOs were working there, food distribution and water were scarce, and shelter conditions were appalling. The contrast between the state of IDP camps and its neighbors, the refugee camps, was striking. The Sphere sectors were barely being covered by UN agencies. Assistance for other areas of need, such as education, was almost non-existent.

In December 2006, the Jesuit Refugee Service (JRS) was asked by UNICEF to develop educational programs in the Department of Dar Sila, in the southeastern part of Chad. JRS was specifically requested to open primary schools in the IDP camps surrounding the towns of Goz Beida (Koubigou, Gouroukoun, Koloma, and Gassire) and Koukou-Angarana (Habile 1, 2, and 3 and Aradib) and to support the schools in the villages of Sannour and Abchour, which could be used as camps in case of new waves of displacement. I was the JRS Program Director in the Department of Dar Sila from December 2006 to January 2008. Because of the ongoing war, the social and historical characteristics of the area, and the humanitarian context, providing primary education in IDP camps in eastern Chad proved to be an incredibly challenging task which became a case study for the strengths and weaknesses of the humanitarian response.

Context

Humanitarian emergencies often unfold in places where the situation was already dire. Besides scarce rainfall and chronic instability, eastern Chad has always had one of the lowest schooling rates in Africa.

In 1975, fifteen years after independence, Professor Issa H. Khayar published *Rejecting School: A Contribution to the Study of Education Problems Among the Muslims of Ouaddai*.[1] In it he describes how military resistance against colonial France, and the endurance of the Quranic schools, led to a refusal by a large number of the Ouaddain society (the Department of Dar Sila is the southernmost part of the Ouaddai region) to send their children to "Western schools."

Attempts by the central government, since 1960, to increase the school attendance in eastern Chad have been hindered by an utter lack of financial resources, internal civil wars, centralization of the public services, and the isolation of the Ouaddai region due to a lack of road infrastructure. The historical reluctance of the people in eastern Chad to send their children to school, combined with the incapacity of the government to promote public education, produced a schooling rate in the Department of Dar Sila of only 7–10 percent before the IDP crisis.[2]

The crisis only made things worse. By early 2007 almost 120,000 people were displaced, most of them living in IDP camps. According to the representatives of the Department of Education in Goz Beida, there had been 132 primary schools in Dar Sila in 2005. Even if these figures are clearly unrealistic, it is true that there were a small number of schools and government teachers in the region.[3] When violence erupted these teachers were the first to flee, and schools and villages were burned. By December 2006 there were only four functioning primary schools in the whole of Dar Sila. The only secondary school deserving that name was in Goz Beida. In Koukou-Angarana there was a second school with eight students and one teacher.

The massive displacement of people is a drama that knows few positive outcomes. Poor Chadian peasants in the Department of Dar Sila were forced to abandon their villages and settle in IDP sites without enough water or food. In the middle of human catastrophes, new patterns and behaviors sometimes emerge that can improve the life of societies. In the many interviews we conducted upon our arrival to the IDP sites, women and men identified the lack of education as not only one of their most urgent needs but also as one of the main causes of their situation. We realized that a people that had been reluctant or indifferent to schooling for a century were changing their views.

Providing education in this context proved to be a daunting task, but the will of the people made it possible. I will run through the main stumbling blocks we found in our way, examining how we tried to overcome them.

Culture and Context of Displacement

While NGOs and UN agencies awake to the reality of a region after a crisis has erupted, communities suffer problems (and have resources) long before the arrival of our vehicles. The lack of a schooling tradition was the single most problematic reality that we faced when setting up educational programs for IDPs in Dar Sila. Each IDP site in Dar Sila was formed by people coming from at least twenty different villages. During our first assessment we discovered that in some of these sites, such as Aradib, none of the villages represented has ever had a school. As described by a Chadian government official who was visiting the region, "the population was virgin of formal education." This meant that even after a school was opened and attendance in the first days was high, often due to distribution of educational material, it was going to take time until parents and children understood and accepted the fact that going to school is a daily obligation and should be seen as normal an activity as eating. This is only self-evident in other communities where school has accompanied life for decades. Creating this new habit was the first objective of our work.

This could only be done with the commitment and will of IDP families for whom sending their children to school meant giving up a source of income which was sometimes crucial for the family. In eastern Chad, as in so many other developing countries, children are an essential part of the economic life of the family, either directly by working on the fields and selling goods in the market, or indirectly by cooking and taking care of children, thus allowing time for the mother to undertake other livelihood activities. This is the daily reality of rural societies all over Africa and displacement severely aggravates the situation by overstretching subsistence resources.

In the first three months after the opening of the IDP sites around the towns of Goz Beida and Koukou-Angarana, WFP distributed food only once. Since then, the regularity and frequency of food distribution for IDPs have barely improved. The first thing we saw when we drove for the first time toward the IDP site of Koloma, two kilometers away from the town of Goz Beida, were scores of children walking toward the town to fetch water or sell goods in the market. As much as we have to fight to improve the material conditions of displaced families in emergencies, it is obvious that in the context of poverty and displacement only the determined will of the community, never the imposition of an NGO, can make an educational program successful.

There are other cultural realities that can seriously hinder the success of schooling projects in emergencies. Early marriages of girls are customary among many groups in eastern Chad. Girls as young as twelve or thirteen are being prepared for marriage. Some parents believe that a girl who is soon to become a wife and a mother should not waste time learning how to read and write instead of learning the household chores. In these cases, as in so many others, humanitarian agencies are faced with the limits of our interventions: the debate about our capacity—or right—to change cultures is urgent and complex. In most the cases, like in eastern Chad, we can only support the schooling of girls in the system through awareness-raising, in the wider sense, and we can implement projects such as the creation of nurseries in each school. JRS has been especially interested in training the teachers on gender issues and in encouraging the training of female teachers to work in IDP schools. Female teachers are role models for girls, opening their minds to new possibilities for their lives.

Teachers

Teachers are the core of any educational system. Schools can be built, educational materials can be distributed and the availability of children is endless, but teachers are chronically hard to find, and good teachers even more so. This is the reason why any policy or initiative regarding education in emergencies should focus on teachers. It is a worldwide reality that teachers are underpaid and a dramatic fact in the developing world that there are not enough of them.[4] If the life of a teacher is never an easy one, humanitarian emergencies take this situation to the extreme. The number of teachers available can decrease for two main reasons, the context of violence and the sudden arrival of international UN agencies and NGOs.

Teachers are generally the most educated segment of rural communities. In many cases, as in eastern Chad, they are deployed by the Ministry of Education. Because they are usually not a native of the village where they work, they are often the first to flee when violence erupts. However, unlike the rest of the community, they do not flee to IDP camps but to their hometowns or to the capital. Less obvious but much more disquieting for us is the role that UN agencies and NGOs play in the disappearance of teachers from the classrooms. When humanitarian organizations parachute into emergencies, our first priority is to find capable local staff and translators as soon as possible; and teachers are often the only people who can speak both the local and the colonial language. In the Department of Dar Sila there was not a single UN agency or NGO that had not hired one, or several, former teachers. There is not much that can be done about this, as teachers have the right to work for a better salary. However, it is essential to understand the side effects of humanitarian interventions and the harm that they often cause.

The sudden arrival of humanitarian organizations in places like Dar Sila not only pulls teachers out of classrooms, but it also encourages them to aspire to work elsewhere. According to Chadian Ministry of Education rules, community teachers, those who have not followed the official national training, should be paid by the parents. In 2007 the parents had to pay 25 CFA francs per month, a sum difficult for them to attain. A cook or a night watchman working for an NGO or a UN agency would earn no less than 75,000 CFA francs per month. Needless to say, humanitarian jobs were seen as much better deals than being a schoolteacher. The economic situation of teachers in the Department Dar Sila suddenly deteriorated with the arrival of the NGOs, not only in comparative but also in absolute terms. The skyrocketing of salaries paid by the humanitarian organizations caused the immediate increase of prices in the markets of Goz Beida and Koukou-Angarana, making it more difficult for teachers to provide food for their families. This led to a constant, and understandable, request for pay raises by teachers.

The question of who should pay for the salaries of teachers during humanitarian emergencies continues to be one of the most controversial subjects among agencies working in this field. UNICEF has always been reluctant to pay teachers in IDP sites. Although JRS supports the principle that it is the obligation of the state to pay its teachers, the humanitarian crisis in eastern Chad is a clear example of how principles can sometimes endanger the basic rights of people. The Chadian government has been historically unable to assure the payment of teachers in the country. The ongoing civil war that has plagued eastern Chad for the last three years has further crippled the capacity of the central government to pay the few teachers deployed to the region. The Ministry of Education, aware of this reality, put in the hands of the parents the charge of providing a salary to the teachers of their children. If under normal circumstances villagers were barely able to earn the minimum amount needed to pay for community teachers, they were completely unable to do so after displacement, when they had no arable land available in IDP camps and had become dependent on unreliable food distribution. The practice of user fees in education is becoming discredited worldwide.[5]

When JRS started opening schools in the IDP sites, shoulder to shoulder with the communities, the choice was clear: either we would provide wages for the teachers or nobody else would pay them. The State was absent, and the parents were unable to do so. We thought that the right of the children to receive an education and the positive impact that even some years of schooling have in the life of people came first.[6] We gave the teachers incentives, wages that amounted to what the communities, under normal circumstances, would be paying them. With prices distorted by the humanitarian world and colleagues earning three times more doing low-skilled jobs, this was not much, but it made it possible for schools to open and run. The payment of incentives was agreed upon by the Department of Education of Dar Sila, representing the State, and the parents' association of each school. The agreement made it clear that the financial support would cease when the displacement was over, since it was understood that a community back in their village would manage to pay for its teachers. The opening of the schools, in parallel, allowed us to lobby the government to send teachers back to the region and by October 2008, new teachers arrived, sent from the capital, N'djamena. Their number was small, but the principle was set.

The second big question concerning the teachers in an emergency situation, apart from the salaries, is the training. Too often the humanitarian world, often shortsighted with its one-year projects and the unwillingness of donors to long-term commitments, has followed the shortcut of offering teacher-training programs that generally last between one week and three months. From our point of view, these programs have been one of the most important failures of the educational initiatives that NGOs and UNICEF have undertaken in emergencies. Short teacher-training programs are easy to organize and cost less, and the photos always look nice in the annual reports. On the other hand, their ability to change the capacity of teachers to improve their educational skills is low. In most of the countries where emergencies take place teachers do not only lack an understanding of modern and participative ways of handling a classroom, they lack the basic knowledge required to teach the pupils each subject. In the Department of Dar Sila, this reality was strikingly obvious and painfully widespread. According to the local authorities, the illiteracy rate among adults in Dar Sila, before the IDP crisis, was 85 percent. Those who knew how to read and write were few and their knowledge of any of the teaching languages (in Chad, primary education can be received in Arabic and French) was small. Previous teacher-training programs had been unsuccessful in improving the capacity of the teachers and in retaining them in the school system.

Initial assessments showed that some of the community teachers that were giving classes in the IDP sites had not even finished their primary education and that only one of them out of the first ninety-four had finished secondary education. This meant that they were teaching subjects they had not mastered. After long meetings with the parents and the Department of Education, we decided to start an ambitious teacher-training program. The teachers were to be trained in order to pass the BEPC official exam (intermediate secondary). This would give them a diploma that would allow them to officially enter the national education system. This implied a long-term project and a long-term commitment because the teachers would have to be trained for no less than a year. Since the beginning of the project in September 2007, the teachers have been enthusiastic to make this commitment.

Teacher-training programs confront the dual reality of education in emergencies. Many of us consider education a life-saving sector—especially in the case of girls, for, according to the World Health Organization, each year of a girl's schooling equates to a 5–10 percent reduction in infant death—and we strongly support its establishment in the first stages of a humanitarian response. But, at the same time, sustainability is fundamental to any educational program, and without it the project is a failure. There are no quick wins in education. Short teacher-training programs are often no more than good-looking patches instead of useful activities. Sustainability, even in emergencies, should be our main objective. We have to think about what communities will have when they return in terms of material and resources, and we have to undertake programs that will enhance the capacity of teachers and the capacity of communities themselves, especially through the parents' associations.

Parents' Associations

The Ministry of Education in Chad recognizes the Associations des Parents d'Elèves (parents' associations) as the institutions in charge of managing community schools. Through them, community teachers are recruited and paid. In the Department of Dar Sila this aim is confronted by the absence of a school tradition. When we started opening primary schools in the IDP sites of Koubigou, Gouroukoun, Koloma, Gassire, Habile 1, Habile 2, Habile 3, Aradib 1, and Aradib 2, most of the parents elected to become part of the parents' association had never belonged to one and had never been to school. The election itself of the parents' association in the IDP site schools of Dar Sila was remarkably complicated for several reasons: no less than twenty villages are present in each site, often coming from different geographical areas and sometimes with serious problems of understanding among each other. It is difficult to involve people in a school that belongs to so many actors. Making each village feel not only just a part but also an owner of the school, and therefore responsible for it, proved to be a challenging task.

It is nonetheless essential. As the humanitarian world gets increasingly professionalized and donors request time-bound and achievable projects that have to be measured against a given deadline, one of the first casualties tends to be the beneficiaries' participation.[7] Beneficiaries' participation, as much as sustainability, should constitute the backbone of any project proposal concerning education in emergencies; not only on paper but also in real life. Unfortunately too often it is easier for NGOs and UN agencies to push programs on putting money over the table than to get tangled in meetings with IDP or refugee leaders who are not always eager to work for the projects they benefit from. It is important to understand that this is the way of creating dependence and undermining ownership.

While helping communities to pay for the teachers through incentives, JRS urged the parents' associations to fulfill their obligations and become the real managers of the IDP site schools. They were in charge of building the classrooms where needed and rehabilitated the hangars that UNICEF had built with local materials. The parents' associations also took part in the distribution of educational material and in the school feeding programs. This participation has not always been easy, due to disagreements among the several villages and punctual cases of corruption, but these difficulties have been outnumbered by the response of the communities and their involvement in the education of their children.

From the very start, parents' associations followed specific training programs in order to enhance their ability to run the schools. With the idea of return on the horizon, each village was supposed to have representatives in the parents' association trainings, since each village will have to replicate the same model when its people get back home. As much as the context of displacement increases the complexities of developing an educational project, local culture, as we have seen before, may also become an issue. According to the Chadian Ministry of Education rules, the parents' associations have to be democratically elected and gender-balanced. In a country with no tradition of democracy and in a region alien to the concept of gender balance, it is extremely difficult to achieve these guidelines. Again, international humanitarian organizations have to explore the limits of our intervention. Through awareness-raising programs, communities were informed of the rules set up by their own government, and the Dar Sila departmental inspectorate for national education firmly urged the leaders to follow them. Despite these initiatives, it was very uncommon to see women attending the parents' associations meetings, yet even in that context more and more women started to get involved in school management.

The parents' associations are an extremely powerful tool of cohesion in an IDP site. The different communities are forced to work together in order to achieve the common goal of having a good school.[8] With one school for all the children it is easy for all to understand that clashes among villages will damage every single child. In the same way that parents' associations are a natural glue for communities living together in an IDP site, the same is valid for the relationship between IDPs and another one of the weakest links in the humanitarian response to a crisis: the host population.

Host Population

During humanitarian emergencies, UN agencies and international NGOs have traditionally neglected host communities. With a limited amount of resources, NGOs and, most tellingly, donor agencies have preferred to focus on refugees and IDPs. The result has been an uneasy relationship between host populations, IDPs, and the humanitarian world that, in some places, have exploded into conflict. Eastern Chad has been an example of this. In the Dar Sila department, conflicts between host populations and IDPs related to jobs, materials distribution, and water have not been uncommon. Education is one of the sectors ripe for conflict. From the start it was decided that all the

IDP sites set up in the outskirts of villages and host villages would share the school. In the few cases where there was a school before the IDPs arrived, such as was the case in Koukou-Angarana, it was expanded and improved. In other cases a school was built for both communities.

It seems clear that the humanitarian world should change its practice regarding host population. Host communities deserve assistance and should profit from humanitarian projects. It is important to build structures that will outlive the camps in order to make the host village understand that something will be left when IDPs, and NGOs, depart. Everything that can be built to last close to the school, like playgrounds, changes the way host communities look at their neighbors.

Parents Associations represent the best space for dialogue and joint work for IDPs and host communities. Tensions were inevitable, especially in villages where during the setting up of the sites conflicts had already emerged like Koubigou or Gassire, but sooner or later all the parents understood that the only victims of their lack of cooperation were their own children.

Children

The only thing we can take for granted in an emergency is the large amount of children in need of education. While normally we tried to provide primary education according to the ages of girls and boys, we soon we found out that many children under five showed up to school with their elder brothers and sisters. Teenagers also want to profit from a school that did not exist when they were kids. This constantly growing number of pupils poses the first key problem that humanitarian organizations dealing with education have to answer after opening a school: what should be the ratio of teacher to pupil. Every Ministry of Education has a document specifying this topic. In many of them, and Chad is a clear case, what is written is no more than a hope, even for schools in peaceful areas.

From January 2007, when we recruited the first teachers and started opening schools, it was obvious that there would never be enough teachers to provide for classrooms with thirty or forty children. It is impossible to train a teacher in a week and many people would rather take up any job with a UN agency or an NGO rather than being a teacher. The dilemma is simple: either you lower the number of pupils per class—improving, in theory, the quality of education—or you overcrowd the classes, thus avoiding having children in the streets. For JRS in the Department of Dar Sila the choice was plain. The first priority was to create a school tradition in an area where it did not exist. We wanted, and the parents wanted, to have every child in school. Secondly, eastern Chad is an area where forced child recruitment has been rife, especially since March 2007. The protection that a school gives against recruitment was essential.[9] We therefore tried to reach as many children as possible. It is not an easy decision to make—the quality of education will be affected and the burden on the teacher will increase—but it is the right one. The only answer known to improve the pupil/teacher ratio is to build more classrooms, something relatively easy, and to train more teachers, which is a slow process.

From 2007 to 2009 the openings of schools in IDP sites in the Department of Dar Sila were a success. For the school year 2007–8 the IDP sites schools of Koubigou, Gouroukoun, Koloma, Gassire, Habile 1, Habile 2, Habile 3, Aradib 1, and Aradib 2 and the village schools of Abchour and Sannour were among the very few in the Ouaddai region that opened on the date marked by the Chadian Ministry of Education, October 1. Keeping the children at school was the next step. Cultural patterns, plus the context of displacement, make schools somehow unnatural places for a child to be in the IDP sites of eastern Chad. There are external circumstances that prevent children from attending school. The most important one is hunger. Our statistics showed that as time passed on from the food distribution date in the sites, more and more children were not reaching the classrooms in the morning; in some cases, as much as three months elapsed between two food distributions. It was common during the rainy season—when the *wadis* are filled with water—to see children going to fish instead of going to school. It is impossible to ask a child who has not eaten in the morning to go learn the alphabet.

We were in favor of frequent and regular food distributions but, more successfully, we held meetings with WFP in Abéché to commence school feeding programs in the IDP sites of Dar Sila. In January 2009, the NGO Feed the Children launched the school feeding program, and the results were immediate. School attendance grew and, what is more relevant, remained stable. Too often school feeding programs have been considered development projects and therefore have not been associated with education in emergencies. School feeding programs should be just as essential in education programs as the distribution of educational material. Filling classes is simple; keeping children in them is much more difficult.

There are also internal circumstances that may push children away from school. It is essential that the child feel that he or she, individually, is important in the classroom and that his or her absence will be noticed. To achieve this we used two methods. Teachers were urged to keep attendance records. According to the Chadian Ministry of Education teachers are obliged to do so, but the reality is that all over the country it is unusual to find community schools where teachers call the pupils' names daily. It was not easy, due to the number of children in each class, but it was necessary. The second initiative we took was to create a team of community workers trained to follow the cases of absenteeism. These teams would check the attendance lists provided by the teachers to find out which children had been repeatedly absent and discover the cause, normally trying to reach the family. It is a hard task because of the number of children but, once again, it creates the feeling that there is a difference between showing up and not. It is crucial to remember that throughout the implementation of the program we have emphasized that, in a context like Dar Sila, the priority is to make the school a common part of the community's life.

Schooling Versus Education

Once children are in classroom, and a teacher in front of them, the true challenge is education. One of the biggest insincerities of NGOs and UN agencies working in this field is calling education what is in fact schooling. The MDGs on Education and the Education for All

Objectives are not free from this misconception. Schooling is, in itself, a right and an achievement. But it should not be called education.

Education is a process that requires time, and NGOs and UN agencies working in emergencies should be humble enough to understand that achieving it is generally beyond their capacity. Education requires that the level of teachers is high enough to permit a child to climb up the ladder of higher education, that the state is capable of providing children with a chance to improve their life by studying hard, and that society creates an atmosphere where literacy can be encouraged and sustained through the existence of books (and, ideally, computers). In theory, this is simple to achieve. The quality of teachers training programs should be improved, the educational system of a country should be enlarged and children and adult literacy should be promoted. Apart from the advocacy work that those working on education in emergencies have to undertake, we can and should also have a lasting impact at the field level.

I have mentioned the importance of carrying out teacher training programs for IDP schoolteachers that solidly improve their teaching capacity and knowledge. At the same time, shortly after the opening of primary education schools in emergency settings, adult literacy centers should also be offered to the community. In Dar Sila, creating adult literacy programs faced the same problem as creating schools: there were not enough teachers. Good teacher-training programs are the only way to leave this vicious circle.

We can also assist in improving the capacity of the local authorities. Working together with the representatives of the Ministry of Education is not only an obligation—it is their country, they authorize the existence of the IDP schools and validate the marks of children at year end—it is also an opportunity to improve our activities and their skills. Historically the relationship between the rushed humanitarian world and the local authorities has been complicated, and sometimes NGOs have tried to get around them; we tend to forget that we will leave while they will stay. However, if a school is to last and succeed in giving children an education that helps them open new possibilities in life, the state has to be an essential part of that process. After our arrival in eastern Chad, we worked closely with the Department of

Education in Goz Beida and Koukou-Angarana. They were involved in every step of the process that led to the opening and functioning of the IDP sites schools. Whenever there was a conflict, its intervention was instrumental in assisting teachers understand how the humanitarian world works and our relationship with the state.

Conclusion

Education is the goal, schooling the first rung. Universal education is achievable only if the state is committed.[10] Since 2007 the displaced children of Dar Sila have schools to attend in the middle of a dramatic humanitarian crisis. At the end of the 2008–2009 school year there were twenty-seven schools opened in the IDP sites and surrounding villages with twenty-seven parents' associations, two hundred teachers, and more than fifteen thousand children studying in their classrooms. If the schooling trend continues, the Department of Education in Goz Beida believes that the MDGs on education are achievable in Dar Sila. In 2009 UNICEF asked JRS to help lead the Education Cluster in eastern Chad. But even if we are satisfied with the work that has been carried out, the challenges remain the same. We must increase the number of children in school and improve the quality of education.

But the problem of these children goes far beyond the IDP sites. Chad is a country where education is almost impossible. The 2009 *Developing Countries School Success Index* of Save the Children ranks how well prepared children are to succeed in school. Chad was last among one hundred countries studied, followed by Afghanistan, Burundi, Guinea-Bissau, and Mali. The emergency in Chad, as in so many other countries, is not only the war-affected region. It is also the lack of progress in providing the society an education that will allow them to rise out of poverty. As much as we consider HIV/AIDS or climate change a global emergency, lack of education should be faced up to as such. The real emergency is not having access to education.

In March 2007, rebels attacked the villages of Tiero and Marena, at the heart of Dar Sila. The families that survived reached Koukou-Angarana after walking under the sun for hours. When they arrived, we took the injured to the hospital and distributed food. When I sat down to talk with the elders I asked them what had happened. They described the horror they had been through. I then asked them why it happened. I was expecting an answer based on politics, tribes, history, or economy. But one of them spoke and said, "This happened because we do not have education." That day we decided together to build the school of Habile 3, and three weeks later it was opened. The amazing will of the people of eastern Chad, their strength to work for their children, and the appreciation that education can transform societies will emerge from this conflict.

Education as a Survival Strategy

Sixty Years of Schooling for Palestinian Refugees

SAM ROSE

In times of war, still photography retains a singular ability to transfix and disturb. From the carefully crafted and sometimes dissembled compositions by the legendary annalists of the American Civil War to the "trophy shots" of torture and abuse of Iraqi detainees in Abu Ghraib, war photography can chronicle the horrors of conflict and the vulnerability of its victims with an often-transcendental power and emotion.

Among the most arresting images from the military operation that Israel waged in Gaza from December 27, 2008, to January 17, 2009, are a series that illustrates both sides of the artistry of the photojournalist. They show the shelling of a UN school in the northern town of Beit Lahiya by the Israeli Army on the last day of the war.[1] In the initial photograph, an Israeli ordnance round can be seen exploding above the school, while another lands in the playground, a plume of white smoke in its wake. In the next, scores of miniature fireballs rain down on a UN compound. Subsequent pictures convey the panic and heroics of those under bombardment—civilians flee toward the camera while paramedics dart in the opposite direction. They disappear into a blanket of dense white smoke, which has now engulfed much of the schoolyard and buildings, where almost two thousand Palestinians had been sheltering. Soon, the paramedics are also in flight, as firefighters struggle in vain to douse the flames.

Suddenly, the mood and pace (and most likely the photographer) changes, and we are taken through several more stylized and sometimes staged photos that portray what is clearly the morning after the attack. Women and children return to the blackened classrooms where they had been sheltering before the bombardment; aid workers and local residents sift though the remains, documenting and coming to terms with what has happened.

We now know that the ordnance used by the Israeli Army that day was filled with felt wedges steeped in white phosphorus. The chemical is fiercely incendiary: it ignites on contact with air and burns at very high temperatures, continuing to do so unless completely deprived of oxygen. It sticks to skin and will destroy tissue, and it also releases highly toxic fumes that can cause serious internal injuries. Although international law permits its use in certain circumstances, it is prohibited wherever there is "a concentration of civilians,"[2] a description that Human Rights Watch has concluded applied on that day in Gaza. We also know that two children, aged five and seven, who were inside the school were killed, and thirteen persons were injured.

The Beit Lahiya Elementary School sustained extensive damages. It is operated by the UN Relief and Works Agency (UNRWA), which provides basic services to around one million Palestine refugees in Gaza, approximately two-thirds of the total population, and another 3.5 million in the West Bank, Jordan, Syria, and Lebanon. Thirty-six of the organization's schools were damaged during IDF Operation Cast Lead, while forty-five were turned into temporary shelters, hosting more than fifty thousand civilians forced to flee their homes.

This is not the first time that UNRWA installations have been damaged or destroyed during military operations. The organization's entire compound in Nahr el Bared camp in Lebanon, which housed five schools and several other UN and NGO facilities, was leveled by the Lebanese Army in the siege of summer 2007. Ninety of UNRWA's one hundred schools in Gaza were damaged or looted during the 1967 Arab-Israel war. The almost serial nature of such incidents and the repeated breaches of the immunity and inviolability of UN assets that they imply should in no way reduce our shock and duty to protest in the strongest possible terms.

IDF Operation Cast Lead ended on January 17, 2009. A week later, on January 24, following makeshift repairs, UNRWA reopened the Beit Lahiya school. Indeed, UNRWA resumed schooling in all its schools in Gaza, reaching close to 200,000 refugee pupils. This was arguably the most significant step taken by UNRWA in the immediate aftermath of the war. In addition to creating space for children and parents alike—a precious commodity in the teeming sprawl of Gaza—to come to terms with their trauma, it helped to provide structure and stability for young people whose lives had been shattered by war, reinforcing a return to "normality," insofar as such a word has meaning in Gaza these days. It also had significant symbolic value, emphasizing UNRWA's desire to reorient itself rapidly toward longer-term development goals and its commitment to supporting the Palestine refugees of Gaza during the long recovery process.

The Early History of UNRWA's Education Programs

Behind the incident described above lies the story of one of the largest education systems in the Middle East and one of the most important in contemporary Palestinian society.

While long-term crises and sudden shocks in the past decade have forced UNRWA to direct an increasing share of its resources to emergency relief, over the course of its sixty-year history the organization's primary role has been to provide social services—mainly education and health—to an ever-growing population of refugees. Each day of the academic year, 500,000 children aged between five and fourteen pass through the doors of 689 UNRWA schools in Jordan, Lebanon, Syria, the West Bank, and the Gaza Strip, at an annual cost of more than US$280 million.[3] Since operations began, UNRWA has educated three generations of refugees, or around four million children.

It is in the sphere of education that UNRWA has made by far its most significant contribution to Palestinian human development. However, this was not the original intention of the UN when it began to seek funds for Palestine refugees in the wake of the 1948 Arab-Israeli conflict. In what was thought would be a temporary emergency operation, no provision was made for education by the UN's Disaster Relief Project for the victims of the war, which began work in July 1948, or in the first budget of its successor, UN Relief for Palestine

Refugees (UNRPR).[4] The few schools there were in the refugee camps that had sprung up across the region had been set up by voluntary organizations, including the Quakers in Gaza and the Red Cross Society in the West Bank, and were supported by UNESCO.[5] They survived on sales of used materials, special appeals and grants from other UN organizations.[6]

When UNRWA assumed responsibility for the education of Palestine refugees in the middle of 1950, it took over sixty-one schools from the UNRPR, with 730 teachers and 33,600 pupils, representing just over a quarter of all school aged refugee children at that time.[7] A director was assigned from UNESCO to lead the agency's education programs, in an arrangement that continues to this day. However, the share of education in the organization's first budgets was negligible: between 1950 and 1955, it accounted for only 5 percent of total expenditures.[8] While understandable given the context—more than 700,000 refugees had recently been uprooted from their homes and livelihoods—it also reflected the prevailing approach of the major powers to the Palestine refugee "problem."

Unable to repatriate the refugees as per the terms of UN General Assembly Resolution 194 (III) of December 11, 1948, the international community instead sought their regional resettlement and "reintegration." This was to be achieved through public works projects that would (1) help displaced communities to become self-sufficient; (2) strengthen the economies of the host countries; and (3) allow UNRWA to delete large numbers from its ration rolls in fairly short order.[9] Projects developed to meet these goals varied in scale from local construction works, for example, of shelters, roads and camps, to regional initiatives of a scope and ambition that have not been seen since in more than sixty years of peace-building efforts for Palestinians and Israelis.[10]

UNRWA had already begun increasing its expenditures on education, partly to support the goal of economic rehabilitation, but also given the longer-term character the refugee crisis was gradually assuming. Another consideration may have been a rebuke from UNESCO's Executive Board to the effect that the UN would be in breach of the Universal Declaration of Human Rights if it failed to make further investments in educating Palestine refugee children.[11]

Although food parcels still accounted for the majority of UNRWA's budgets, there were quantitative leaps in allocations to education during the first half of the 1950s. By the middle of the decade, its total budget share had increased to almost one-fifth of total expenditures, a twentyfold increase since 1952. Almost 100,000 pupils were being taught in 264 UNRWA schools by 2,700 teachers, while UNRWA funding subsidized the education of another 55,000 refugee children at government or private schools.

Most teaching had moved out of tents and into specially constructed premises or rented buildings. Indeed, UNRWA schools were the first permanent structures in many camps and became a symbol of an increasingly education-centered approach to refugee assistance. UNRWA was also operating two vocational training centers in Qalandia and Gaza and had links with regional organizations and host governments to secure placements and employment contracts for its graduates at the end of their courses. Course materials were developed with regional market needs—and UNRWA-funded large-scale development projects—in mind, and with the support of specialist UN agencies. University scholarship and teacher-training programs were also offered, albeit on a very small scale, while tens of thousands of adult refugees benefited from adult literacy ("fundamental education") courses, part of a campaign to combat illiteracy.

As the 1950s progressed, skepticism about the feasibility of the much-trumped schemes of economic reintegration and resettlement continued to grow. The failure to resolve the Arab-Israeli conflict as per the terms of Resolution 194, the intransigence of Israel and host authorities, and the paucity of donor funding for development plans were all contributing factors. Perhaps most importantly, the refugees refused to be part of what they considered to be a conspiracy to liquidate them and deny them their internationally recognized rights.

The Suez crisis of 1956 signaled the abrupt end of the era of regional refugee-development schemes. After this time and the final acceptance that there could be no resolution of the Palestine refugee problem absent political accommodation, UNRWA began to reorient itself away from relief toward education. The seeds of this strategic transformation had been planted in the five-year education plan of 1955–1959 and were given full voice in the watershed annual report of 1960.[12] In that report, UNRWA's fourth director, John Davis, announced a three-year plan for "well planned and promptly executed programmers for improving general education and expanding specialized types of training" that would harvest the "latent productive talents" of refugee youth that had so far been laid to waste.[13]

Proposals included the formalization of a three-year lower preparatory cycle; increasing the number of vocational trainees from 500 to between 2,000 and 2,500 through the construction of six new training centers and the establishment of a joint vocational and teacher-training school for girls and a separate men's teacher training center; doubling the number of university scholarships awarded by UNRWA each year, from 90 to 180 per year; and institutionalizing UNRWA's loans and grants programs, designed to help refugees become self-sufficient.[14]

These initiatives, which were revolutionary in the Near East of the 1960s, laid the foundations for the remarkable successes of UNRWA's education programs in the years that followed. They established UNRWA as a significant educational force, far ahead of the curve in the Middle East and North Africa region, and creating what Israeli sociologist and anthropologist Maya Rosenfeld has referred to as an "educational advantage" for refugees over their non-refugee peers in the Arab world, which persisted for many decades.[15] To this day, literacy rates of Palestinian refugees are higher than for the Middle East and North Africa region as a whole, particularly among females.[16] UNRWA's initiatives were often pioneering, especially in girls' education and vocational training, where graduates were highly sought after and an UNRWA qualification effectively guaranteed employment. The organization had a reputation for innovation, dynamism, and vision in its approach and commitment to education, a commitment that was shared by the refugees.

Beginning in 1961, the share of the budget spent on education began to increase significantly, and by 1970 it had become UNRWA's largest program. Five years later, education accounted for more than half of all program expenditures.[17]

The Successes of UNRWA's Education Program

The successes of the program can be attributed to a number of factors. Some were circumstantial, others a result of direct choices and support by UNRWA, the host authorities and the refugees. Considered as a whole, they provide compelling evidence of the opportunities for development and progress that can emerge from crisis and tragedy.

The First Decade

Even before the Palestinian exodus or *nakba* of 1948, education levels amongst Arabs in mandate Palestine were relatively good by regional standards, despite Palestinians' primarily agrarian backgrounds. While in no small part due to investments by the British in the last years of the mandate period, this also reflected a growing desire for learning among village communities at that time, with many contributing labor to build their own schools and the government providing the teachers.[18] Nevertheless, overall attainment levels were quite low, and enrollment was far from universal. Schools were still based predominantly in towns, limiting access for the majority rural population and particularly affecting girls, who fell victim to cultural barriers regarding unaccompanied travel.[19] Estimates of illiteracy among Palestinians at that time range from between 60 and 70 percent, with around half of all children—and four in five girls—not receiving any kind of education.[20]

The *nakba* removed many of the barriers to physical and cultural access, particularly for rural Palestinians, who formed the overwhelming majority of the residents of the refugee camps. The network of schools established by voluntary organizations offered a platform that UNRWA was able to capitalize on and support with the resources, expertise, and administrative structures that they needed to flourish. Within months, many thousands of Palestinians now had educational establishments within walking distance. Enrollments by girls increased with each year, suggestive of how cultural norms that might conspire against progress in normal circumstances can at times be rapidly overcome in times of crisis.

If UNRWA's attitude to education in the 1950s could be seen as somewhat ambivalent, the same could not be said of the refugees. As early as 1951 refugee communities were urging UNRWA to devote more of its resources to education,[21] and while often dubious about its overall intentions and those of its powerbrokers, they viewed education as an apolitical form of rehabilitation, which did not compromise their right of return. UNRWA's commitment to education from 1960 onward was also framed explicitly in these terms.[22]

For the first generation of Palestine refugees, education was both a short-term and long-term coping strategy. As a result of the *nakba*, a predominantly peasant society had lost access to its land and needed to secure alternative and reliable means of livelihood. Education and training were viewed as well suited to their needs and capacities; the human capital that they generated could not be destroyed or taken away.[23] However sepia-tinged this analysis may appear, there is no doubt that study offered a route out of poverty, and as such, investment in education—particularly higher education—was embraced as a "family project"[24] by many households. Some researchers have further suggested that refugees' desire to better themselves partly reflected a sense of inferiority next to Israelis. Bridging the gap thus became part of the national struggle, one in which education partially replaced the homeland as a focus for Palestinian efforts.[25] There were also more mundane reasons for this increased interest: children who had previously worked in the fields no longer had the opportunity to do so and needed alternative ways to occupy their time.

The 1960s and Beyond: The Impact of Davis's Reforms

Davis's plan yielded almost immediate results. It fostered the growth of what could be described as a virtuous circle for education similar to those seen in many western European states following the Second

World War, and which acted as a vital engine for economic growth in the postwar period.

The institutionalization of a third year of preparatory education, which effectively formalized the lower secondary cycle in UNRWA schools and with it nine years of compulsory education, was critical. This facilitated the continuation of refugee students into upper secondary and also tertiary education, which was often provided free of charge by the host authorities and Eastern Bloc countries.[26] By the end of the 1960s, the ratio of Palestinian university students to total Palestinian population was considerably higher than for the Arab region as a whole and comparable to many developed countries. The fact that most of these students came from families of peasants who would have been fortunate to have access to even basic education a generation earlier makes this achievement even more impressive. As was the case with primary education, girls in particular benefited, and host country data show major leaps in refugee females' participation in higher education from this point.

The expansion of UNRWA's education and training programs enabled refugees to tap into the professional opportunities offered by the opening up of regional markets, particularly in the Gulf. Oil-rich governments were making massive investments at that time, including in schools and hospitals, and UNRWA was quick to pinpoint these states as potential employers for graduates. Market surveys and systematic analysis of labor market needs across the Middle East, played an important role, continually informing the content of training courses at the new vocational colleges.[27] UNRWA's own employment placement centers, originally set up to facilitate refugee resettlement, provided further support. Investment in education quickly became self-reinforcing, as remissions from graduates of UNRWA schools and colleges working in the Gulf funded the education of future generations of Palestinians from the camps. Importantly, such remissions also helped to support the Palestinian national movement.

Likewise, the development of public institutions and infrastructure in the Arab host countries and UNRWA's own expanding education and health programs ensured a ready stream of work for qualified job seekers, particularly for women, for whom teaching and medical jobs were well suited and culturally accepted, and for graduates of UNRWA technical colleges, who built much of the furniture used in agency schools.

Although partly driven by the failure of attempts to ensure refugee repatriation, UNRWA's decision to reorient itself toward delivery of education services in the 1960s had a transformative effect. It allowed the organization to transcend the temporary nature of its mandate and focus on longer-term human development goals. For the refugees, whose role in shaping UNRWA's priorities should not be underestimated, this move effected a qualitative change in their human capital and living levels.

The Challenges

Achievements in education have come despite chronic underfunding and repeated national and regional emergencies over the past sixty years. The cumulative effect of these has been a dramatic narrowing of refugees' educational advantage and opportunities in recent decades, at a time when new labor force entrants face ever-growing competition for employment in an increasingly global job market.

Funding Shortfalls

UNRWA's dependency on voluntary financial contributions was identified as a major structural shortcoming as early as 1951, in the UNRWA Director's Interim Report of that year, which covered the first five months of operations:

> The Agency's financial situation has never been a happy one. . . . At no time in its brief career has the Agency been able to see its financial position assured for more than a few weeks ahead. A large and complex program cannot be adequately planned and administered efficiently when it depends on the receipt of voluntary contributions in unknown amounts, to be delivered at unknown times.

Financial shortfalls had an immediate impact on UNRWA's education programs, consistently limiting capacity to invest in the human and capital resources needed to accommodate a rapidly growing student population. From the early 1950s UNRWA was forced to operate schools on a "double-shift" basis, with one group of students studying in the morning and a second in the afternoon. Although the planned elimination of such a system became a mantra of Annual Reports, with each passing year the proportion of double-shifted schools grew, and some schools in Lebanon were eventually forced to operate a third shift. By 2004, more than three-quarters of UNRWA's schools operated two shifts each day.[28] As a result, school days had to be shortened to accommodate the second shift, resulting in fewer hours available for learning, use of specialized facilities and extracurricular activities.

Construction plans to reduce class sizes and teacher-pupil ratios and replace rented premises, many of which dated back to the early 1950s and were increasingly ill suited to the demands of modern learning, were prepared and updated but were routinely underresourced. At times, the funding crisis became so grave that it threatened programs' very existence. In the late 1970s, the provision of university scholarships was frozen, while spiraling inflation over the course of that decade led the Commissioner-General to propose the closure of all preparatory schools in 1979. A year later, UNRWA was considering shutting down all its schools for financial reasons.[29]

A Deteriorating External Environment

UNRWA's perennial funding crises and its inability to upgrade and develop its educational resources, facilities, and systems undoubtedly contributed to the eventual narrowing of the educational advantage of Palestine refugees over their Arab peers. Other external factors, which had helped to create such an enabling environment for education throughout the 1960s and 1970s, also began to recede. By the early 1980s, the demand for professional expatriate labor in the Gulf, which had fed the drive for education and whose remittances had supported the education of future generations, was down. The Gulf States were becoming increasingly reliant on their own nationals to fill teaching and other public sector roles, a professional cadre and system whose development owed no small debt to Palestinian migrant workers. At the same time, massive investment in education systems by other Arab States made to absorb their rapidly growing, increasingly urban-dwelling young populations, paid dividends for host-country nationals and led to greater competition for fewer jobs. The same pattern was repeated in the West Bank and Gaza after the establishment of the Palestinian Authority in 1994: donors made significant investments in systems and structures that had been underresourced and heavily neglected by Israel in the years since 1967. Even so, the Gulf States remained an important source of employment for professional Palestinians until the end of the first Gulf War in 1991, when an estimated 300,000 Palestinians were expelled from Kuwait. Particularly for those in the occupied Palestinian territory and Lebanon, there were few domestic alternatives to these professional employment opportunities.

For Palestinians in the occupied Palestinian territory, the Israeli labor market offered a major alternative source of employment. Following the 1967 war, Israel had immediately sought to integrate the Palestinian economy into its own. It incorporated the surplus of low-skilled Palestinian labor—mainly male—into Israeli construction, service, and industrial sectors in jobs that were unattractive to its own workers. Access to relatively well paid job opportunities in Israel from 1967 onward also clearly influenced the attitude of young Palestinian males in Gaza and West Bank to higher education, further reducing the educational dividend for Palestinians.[30]

Perpetual Emergencies

Perhaps most devastating, though, have been the repeated crises across UNRWA's fields, which have forced the closure of schools and led to the suspension of classes and the diversion of funding and attention to more immediate, life-saving concerns. In the Gaza Strip and West Bank particularly, these emergencies have become the norm in recent years. Often political or complex in nature, crises have had both immediate and long-term, irrevocable impacts on service delivery, creating new dynamics and realities that UNRWA and its pupils have been forced to adapt to and negotiate.

The Arab-Israeli war of June 1967 is a case in point. Some 274 UNRWA teachers were outside Gaza at the time of the war and were not permitted by the new occupying power Israel to return in time for the start of the next school year. More significantly, Israel exerted its authority to ban the entry of most of the seventy-nine Egyptian textbooks used in UNRWA schools in Gaza. Israel had long been critical of these books, which in its view offered a "distorted account of the events leading up to and following the establishment of the State of Israel and tended to induce hatred of Israel in the minds of the children using them." In coordination with UNESCO, UNRWA was forced to produce its own temporary textbooks and teaching notes.

Special arrangements also had to be made to allow several thousands of students to complete secondary-school leaving examinations (*tawjihi*), a requirement for continuing education in the Arab world. Some students were permitted to travel to Jordan to complete their studies, while UNRWA and UNESCO succeeded in bringing examination papers into Gaza for the remainder. This entailed a complex logistical effort, involving a team of forty international staff, as well as hundreds of inspectors and teachers, and representatives of the ICRC and United Nations Emergency Force. These arrangements remained in place for twelve years, until the signing of a peace deal between Israel and Egypt in 1979.[31]

Two Decades of Crisis in Gaza and the West Bank

The eruption of the first Intifada in Gaza in 1988 effectively marked the start of two decades of upheaval and crisis for the education system in the occupied Palestinian territory, which continue to this day. Entire generations of refugee children and youth have been schooled during this period, which has seen a dramatic narrowing of refugees' educational advantage and opportunities for progress through learning.

The education system was directly targeted during the first Intifada, and schools and students were often themselves involved in violent confrontations with the occupying power. In Gaza, UNRWA schools were closed between 35 and 50 percent of the time, with even greater amounts of time lost in the West Bank, where UNRWA's three training centers were also shut for more than two years from the spring of 1988.[32]

Although students and youth have been less directly involved in the qualitatively more violent second Intifada, its continuing impact on the education process has been no less devastating. Since September 2000, Israel has significantly tightened its system of restrictions on the movement of Palestinians and Palestinian goods into, out of, and within the West Bank and Gaza, with dramatic consequences for all aspects of Palestinian life.[33] The right to education for many has been systematically undermined and violated, with movement of teachers and pupils constrained by checkpoints and closure and regulated by Israeli-issued permits, which are routinely denied. Since 2004, Israel has banned Palestinian students in Gaza from studying at institutes of higher education in the West Bank and Israel. Palestinians in Gaza seeking to travel abroad to pursue disciplines not available in Gaza have also been prevented from leaving, including a number who had been awarded Fulbright scholarships to study in the United States.[34] In the West Bank, the national character of many universities is being eroded by travel and movement restrictions.

Levels of violence have spiraled, including both isolated spikes, for example, raids into refugee camps or targeted assassinations of persons wanted by Israel, and more protracted military operations such as the recent war on Gaza, Operation Defensive Shield in the West Bank in 2002, and inter-Palestinian clashes in Gaza over the past two years. As well as direct physical and material losses—313 children were killed in Gaza during Operation Cast Lead, according to the Palestinian Centre for Human Rights, and 181 schools partially or totally destroyed[35]—these exact high social and psychological costs for pupils, teachers and families, with deleterious immediate and longer-term consequences.

Recent years have also seen an escalation of attacks on the Palestinian national curriculum, whose introduction coincided with the start of the second Intifada. It its most legitimate form, scrutiny of the curriculum has helped to promote constructive debate on the politics of textbooks, including the appropriate framing of history for Palestinian schoolchildren and the role of education in promoting peace and shaping national identity. However, the focus of debate has often moved beyond purely scholarly considerations, with various pressure groups framing Palestinian textbooks as a source of conflict, designed to promote violence and incite hatred of Israel. The allegations and accompanying advocacy campaigns have been used to discredit the Palestinian Authority (PA) and also formed part of an effort to undermine international support for UNRWA. This is despite the fact that independent research has concluded that allegations are based on claims that are "tendentious and highly misleading," and that, notwithstanding some elements of imbalance, the overall orientation of the Palestinian curriculum is "peaceful, despite the harsh and violent realities on the ground."[36]

In some cases, education providers have been able to negotiate their way around some of the physical and administrative obstacles created by the second Intifada, or at least minimize their effects. This includes the reassignment of teachers to schools nearer their homes, the introduction of remedial learning and recreational programmers, and the provision of free stationery and meals to poor students.[37]

Efforts to improve the quality of education have also continued. The PA has persisted with an ambitious program of reform, with the World Bank recently commending the remarkable advances made since the start of the second Intifada.[38] Despite campaigns by its detractors, independent research has also highlighted the constructive and systematic steps taken by the PA to constantly reform and improve the curriculum, including focusing on faculties of critical thinking and creative thinking and promoting concepts of human rights, democracy, and pluralism. For its part, UNRWA has introduced a human rights, conflict resolution, and tolerance curriculum in its schools, integrated into lesson plans, with children as young as five learning the principles, history, and core values of international human rights instruments.

There are also indications that Palestinian families have developed a fairly sophisticated armory of coping strategies to adapt to the new reality, reflecting a continuing, even heightened commitment to education over the past decade. Data from the Palestinian Central Bureau of Statistics show that educational attendance among both males and females has increased at every stage of the education process in Gaza and West Bank since 2000, despite the decreasing ability of parents to support their children's higher education. Research has suggested links between increased investment in education and the loss of unskilled wage labor opportunities in Israel due to closures.[39]

At the same time, the protracted crisis in the occupied Palestinian territory is also clearly influencing the resources and attention that service providers can devote to education, given other competing priorities. Over the past decade, donor funding has been increasingly directed toward humanitarian relief and—in the case of the PA—the budget support that it needs to ensure its survival, at the expense of more strategic, longer-term, and productive development investment. UNRWA expenditure data in the West Bank, for example, indicates that more has been spent in that field on relief activities—including food aid, cash assistance and temporary employment programmers—than education since 2002.

In addition, recent testing in UNRWA schools points to significant drops in educational performance. Although efforts are underway in both Gaza and West Bank to address this through remedial actions and recovery plans, the longer-term consequences of today's protracted crisis on education—those leaving UNRWA schools at the time of writing had barely begun their first grade when the Intifada erupted—will not be fully understood for several years. And while the PA and UNRWA have so far been able to successfully defend themselves against attacks on the integrity of the material used in their classrooms, these campaigns have nonetheless been extremely damaging, particularly in terms of their shaping of the negotiations agenda and diverting attention away from evidently more pressing issues.

Conclusion

The history of UNRWA's education system, and that of Palestinian education in general, is one peppered with crises and emergencies that continue to the present day. The attacks on UNRWA schools in Gaza described in the opening paragraphs of this chapter echo similar destruction wrought during the 1967 war; and just as Israel banned the entry of Egyptian textbooks to Gaza in the aftermath of that war, so today it places restrictions on the passage of paper and educational materials, including those concerning UNRWA's human rights curriculum.[40] From direct obstacles such as bans on entry of supplies and enforced closure of schools to indirect hindrances created by closures and checkpoints, the education system itself has often been thrust to the very center of the conflict. This is particularly the case in the occupied Palestinian territory, where the role of education in the formation of national identity, the framing of historical narratives, and the peace process itself remains under intense scrutiny, sometimes for political ends.

Its experiences mirror those of the Palestinian population as a whole over the past sixty years, and indeed "Palestinian education" bears many of the hallmarks of dispossession and statelessness. As the relationship with both the Gulf and Israeli labor markets has shown, investment in higher education has typically been geared less toward national Palestinian economic needs than those of regional markets. The economic fruits of education have been similarly exported, leaving Palestinian families and communities exposed to exogenous and repeated shocks—most recently, the loss of access to Israeli markets and debilitating obstacles to normal economic activity has generated an employment crisis of unprecedented magnitude for Palestinians in the Gaza Strip and West Bank.

In a similar manner, the debate and very content of the Palestinian curriculum is reflective of the complexity of Palestinians' unresolved struggle for statehood and the sometimes uncomfortable relationship between on-the-ground realities and accepted principles of international law. This relationship often borders on the perverse, as shown by the notion of teaching future generations of Palestinians the value and primacy of very rights and laws that the international community denies them on a daily basis.

In focusing on these constraints, challenges, and controversies, we should not allow our attention to be diverted from the transformative role played by education, and UNRWA's programmers in particular, on successive generations of Palestine refugees over the past sixty years. This has ensured social, if not economic, progress that more closely resembles Western European states in the aftermath of the Second World War than that of neighboring Arab states in the immediate postcolonial period. Despite the efforts of pressure groups and detractors to shape public opinion to the contrary, education programmers have also made a significant contribution to stability and state building across the region. Ongoing national Palestinian curriculum reform efforts, as well as UNRWA's dedicated human rights and conflict resolution programs are testament to the vitality of this contribution.

These achievements should not be overlooked by those involved in the latest phase of peace-making diplomacy in the region. As the protracted socioeconomic crisis in the occupied Palestinian territory highlights, it is increasingly clear that any successful political plan must be accompanied with a development and economic plan of the vision and ambition of UNRWA's early years, founded on the successful education system that exists there. This is essential for meeting the future economic needs of Palestinians—both refugees and non-refugees—and ensuring that at times of peace as at war they are able to continue contributing to the productive development of the region.

16 Education in Afghanistan

A Personal Reflection

LESLIE WILSON

After six years working for Save the Children in Afghanistan, I am seldom without an opinion—thoroughly informed or intuited and intensely felt—about the future for Afghan children.[1] This is particularly true with regard to basic health care, education, and protection, to which all Afghan children are entitled no matter how one chooses to characterize the state of affairs in the country: relief, recovery, development, or crisis. Personally, I would categorize Afghanistan's current situation as a man-made emergency that is chronic and socioeconomic as well as political and ideological. Things are made worse, from time to time, by acute and chronic natural phenomena— seasonal flooding, earthquakes, severe winters, and regional drought —against which vulnerable families have few defenses.

Another aspect of the emergency situation in Afghanistan is the long-term neglect of the education sector, which has variously grown, floundered, stagnated, or essentially ceased to be during the past four-plus decades. In considering progress/regress in education, I restrict my observations to what I know best: early childhood development (ECD) and education (ECE) and primary education (grades one to nine), which ensure the foundational rights of children to improve their lives and life prospects. The states of Afghanistan's secondary and higher education warrant reflection, too, since these have a significant impact on Afghanistan's ability to move forward in socioeconomic development and sociopolitical stability within the family of nations.

At the UN General Assembly thematic debate on education in emergencies held in March 2009, Her Highness Sheika Mosah Bint Nasser al Missned commented on "the essentialness of education as the key to ending the many raging wars and conflicts in our world." Her words resonate particularly for Afghanistan as things accelerate in a backward spiral into insecurity. In the current circumstances, if it is possible—and it must be possible—to guarantee children's right to an education during an emergency, whatever its cause. Afghanistan is where the world can make it so.

At that same debate, I noted: "If Afghanistan can keep moving forward, albeit with frustrating and saddening backward and sideward steps along the way, I am sure that all the world's nations can make progress, too, in ensuring that every child and young person's right to an education, even in the midst of crisis, is realized."

Anyone who has spent time in or around the education sector in Afghanistan in recent years can count, and recount, the ways in which education is becoming more accessible.[2] More children are in classrooms. Fledgling parent-teacher associations or school management committees are meeting to improve education quality and bring schools to remote areas. Afghan parents and grandparents proudly discuss their aspirations for children's education.

One amazing and memorable experience puts a bold line under my great hopes for the future of education in Afghanistan. In early 2008, I was visiting classrooms in Shiberghan town in Jawzjan Province of northern Afghanistan. Save the Children was (and still is) an implementing partner for a teacher-training program that focused heavily on experienced teachers and advisers observing teachers' classroom management and methods. The classroom instructor, who still has me smiling to this day, was a woman chemistry teacher. In a long, narrow, dark room with broken down benches and a passel of teenaged boy students, she was teaching about the chemical composition of . . . something. Since I never managed to be a decent Dari speaker, I was not entirely clear on the back and forth between teacher and youngster. But, initially, I was pleased to note that there *was* back and forth, that is, active student participation both answering and asking questions. One student volunteered to draw the compound on the blackboard, and I was pleased to note that there *was* a blackboard, and chalk. He succeeded, and was acknowledged and thanked. Next, an enthusiastic student suggested that he and a few other classmates act out the compound! The teacher agreed, and, much to my amazement and delight, the boys went energetically to the front of the classroom and delivered an applause-inducing lesson for their classmates and observers.

I have been a teacher of American fifth-graders and of Thai children and teens, and have visited many classrooms during my decades in development work. But this Afghan classroom, which was needful—and deserving—of many material things, was without question among the richest in innovation and enthusiasm that I have ever seen. And so one can be hopeful and answer, "Yes!" to the question, "Is progress being made in efforts to improve education for Afghan children and youth?"

Outside that exceptional rural Afghan classroom, much has been made (by the government and others) about the noteworthy acceleration in school enrollment rates among Afghan children following the 2003 Back to School campaign and ongoing efforts to boost school enrollment. In fact, there have been verified and justifiably lauded enrollment gains. Equally laudable is significant enrollment, especially by girls, in nonformal educational initiatives of all kinds, many facilitated or led by non-governmental organizations (NGOs) in partnership with communities in urban neighborhoods and rural villages alike. There is no doubt that millions of Afghan children are learning, formally in institutions, and informally in homes, mosques and, in ever-increasing numbers, in unregulated private schools as well.

What is less discussed is the irrelevance of enrollment if students neither attend school regularly nor advance, based on achievement, from grade to grade.[3] Without question, this remains an enduring issue in Afghanistan. Another essential issue for all government educational initiatives is that of quality: Are students able to learn and develop given the complications the education sector must confront? Afghanistan's severe weather conditions, school learning environments, students' learning motivation and capacity, and teachers' attitudes, techniques, and styles all influence the quality of education and the ability and willingness of children to attend school.

With the amount of attention, care, and funding devoted to the issue of quality by the education ministry, UN agencies, donors and their advisers, for-profit contractors, and Afghan, regional, and international NGOs, the Afghan government can rightly claim some fine examples of progress and showcase programs that are points of pride—in government schools and community/NGO classrooms alike. The burning issue, however, is the significant disparity between the exceptional examples and the more common educational experience of the majority of Afghan schoolchildren.

The disparity primarily has to do with the fact that, in the months and years of Afghanistan's recent rush to reconstruction and development, the education ministries (education and higher education) did not consistently and systematically work in a unified, focused and ministry-directed process with donors, their advisors, relevant UN agencies, NGOs, and contractors.[4] Missing is a national blueprint for what would—and, significantly, would not—constitute a minimum basic package of educational services for Afghan children and youth.[5] Few people, if any, including several successive ministers of education, were ever fully informed about all that donors, advisers and practitioners were doing on any given day vis-à-vis education countrywide.[6]

Curricula for grades one through twelve were developed, vetted, and printed. Yet there was little to no consideration or budget to ensure that the basic curricula could be taught throughout the vast expanse of the country by the existing teacher corps in government schools. Textbooks and supplies were insufficient to meet the tremendous need, as were classrooms and heat for early and long winter days. What good is a curriculum of any quality without qualified teachers, proper texts and decent places to learn?

With its initial failing to take sure and clear control of education sector inputs and activities in 2002, the Afghan government effectively ceded the growth and evolution of educational policy, service development, and delivery to a disparate and not altogether joined-up array of participants, especially for services-delivery by NGOs and contractors.

Parallel tracks of innovation and model initiatives are not inherently negative. NGOs, after all—during decades of absence, ineptitude, indifference, or antipathy—had done admirable and estimable service for many Afghan children by trying to deliver on their right to education. From these programs, including those for Afghan refugee children in Pakistan and Iran, came excellent initiatives that were repatriated to Afghanistan along with its citizens. Important innovations include parent involvement in school management and leadership; student voices that are heard; child-centered teaching and child-friendly classrooms; the introduction of early childhood learning and socialization initiatives; and special attention to girls' education. Additionally, NGOs repatriated a variety of curricula and, in some cases, textbooks and learning materials as well as experienced teachers.[7] All good.

Where the "all good" became problematic was when the variety, disparity, and richness (or lack thereof) did not support the government's need to define and promote a basic package of education services for all Afghan children. Well into 2006, the Ministry of Education, led by political appointees and not professional educators at the highest levels, was, in my estimation, ill advised by too many donors' advisers and other stakeholders. They did not view or assess the educational needs of Afghan children similarly.[8]

Most NGOs chose to continue addressing the educational hopes and dreams of Afghan children and families as they always had: in the communities where they worked, village by village, and according to their own global or regional standards and experience.[9] Few, if any, were compliant with each other. Fewer, for a variety of reasons, including the inability to gain access to them, were following the aforementioned standard curricula uniformly.

The Ministry of Education and various donors tried to move forward with implementing the on-paper core curricula for grades one through twelve and to convince, at least government schools, to use it. Separate initiatives produced (developed, edited, printed, and distributed) textbooks to support the curricula.[10] Between 2002 and 2007, the issues of teachers' qualifications, school management and leadership, classroom size, and school repair and construction protocols were considered and addressed, more or less comprehensively. Together, these issues all highlighted the glaring fact that schools and classrooms did not exist in anywhere near sufficient numbers. Further, some donors' early efforts to build schools according to modern Western standards (at Afghan government insistence, no doubt) were disastrous. The schools were exorbitantly costly and, since local contractors lacking relevant knowledge and skills for building by external standards were engaged for actual construction, they were usually poorly built. Corruption was a factor in the poor quality of construction because all allocated funds clearly were not spent on schools.

All these aspects of education—core curriculum, optimum classroom size, modern teaching and learning materials, teacher qualifications and preparedness, school management and leadership (involving parents, teachers, and ministry officials), and textbooks— are deserving of further scrutiny in Afghanistan. And, thankfully, these and other aspects were finally considered and addressed in a March 2007 Education Sector Strategy,[11] which was spearheaded by then-Minister Haneef Mohammad Atmar. As a result, some ministry specialists and donor-funded advisers sought increasingly to address these issues. Their attention became more focused as they admitted this fact: in well-intended efforts to educate Afghan children over the past seven years—the case of textbooks, for example—much has gone awry.

Currently, all the cited education issues for Afghanistan are likewise challenges for stable, established governments and their ministries in regard to ensuring educational rights. In the Afghanistan of 2009, it is important to remind ourselves of how far the state and status of education for all has evolved, albeit not nearly far enough.

Yet, innumerable times throughout the past six-plus years, I have asked myself and others, "What in the world, in this world, were we expecting?" Think about it: Afghanistan is a country that was really never much of a country by modern nation-state standards, and it is three-plus decades into its own undoing (aided and abetted, undeniably, by a wide variety of outsiders). Now Afghanistan is expected to rebuild itself into a modern representative democracy by and with some of those outsiders who are implicated in the deep past or more recent history of its current state of disruption. In some people's minds, the regrouping and rebuilding was meant to have happened in ten years or less, an impossible task.

Since September 11, 2001, the education of Afghan children, especially girls, has featured emblematically. So, what does education in Afghanistan look like right now? Are things better or worse, for Afghan children and young people seeking to secure their right to a basic education as guaranteed by their constitution and the United Nations Convention on the Rights of the Child?

What is certain is this: The one-step-forward, two-steps-back (or more) nature of progress toward guaranteeing rights and reinforcing hope for Afghan children is unacceptable. Billions of dollars continue to be poured into the country, but the amount of funding is not analogous to the quality and accessibility of education currently offered to Afghan students.

Early international aid efforts, postliberation, focused on many and varied urgent needs in Afghanistan.[12] Millions of dollars, euros, pounds, rupees, yen, and other currencies poured into the country, most via bilateral and multilateral aid agencies. The government established the Afghanistan Reconstruction Trust Fund to hold the billions. Many other mechanisms injected money into the national treasury and budget. Numerous foreign governments infused and used money off budget,[13] and this was a source of great consternation and concern to the Afghan ministries of Foreign Affairs, Finance, and Treasury and keepers of the budget. Still, funds flowed, and efforts to reform and rebuild ministries and public institutions began.

I met many earnest economic advisers, bankers, revenue and taxation specialists, legal systems reformers, and natural resource management experts along the way. Advisers from many nations and national perspectives poured into *soft* ministries: health, education, social services, and women's affairs. Hopes for renewal and reform, if not all-out resurrection, were profound. UN agencies, international contractors, and Afghan, regional, and international NGOs proposed projects with or without funding, and everyone proceeded apace to be part of the change. In the early years, the activity was hopeful and encouraging, if a bit chaotic and of unclear effectiveness or purposefulness. During those years, uncertainty and haphazardness were tolerated, in part, because of the sheer extent of the need across all government sectors and in every community in the land. Now, years later, the continued lack of coordination and results commensurate with investments and donations are simply unconscionable.

The dichotomy of progress versus stalemate is best evidenced in the divergent roads taken by the Ministry of Public Health and the Ministry of Education. The health ministry, in less than a year after development work began in earnest in 2002, came out with its basic package of health services, including facility descriptions (clinics, hospitals) and standards for both infrastructure and staffing. It called for proposals from national and international NGOs willing and able to support existing government facility employees (midwives, nurses, doctors) in delivering basic services according to nationally agreed standards. The ministry was able to issue its call for proposals because it had also secured agreement among major donors to fund the package at specific, rational levels nationwide. Attempts at equitable delivery of essential healthcare services, especially for the most vulnerable Afghans, were and are not without challenges and shortcomings. However, a comprehensive national plan, including a conscious decision to second the delivery of public health services to non-governmental entities, was quickly devised and agreed by the Ministry in consultation with donors and advisers. With uniform, enumerable process, output, and outcome indicators came uniform, rigorous monitoring and evaluation tools and personnel as well as verifiable progress and identified areas for improvement.[14]

No similar process or outcome occurred within the Ministry of Education, though the aforementioned Sector Strategy issued by Minister Atmar was a big step in the right direction. At least the breadth and depth of the situation—and the overwhelming unmet need—were outlined and, on paper, given priority. Steps were taken to name one foreign embassy's development team as the "go to" team over all education reform; a board of overseers (my words) was formed, and a national education forum was held. Sooner rather than later, however, the status quo prevailed. Donors and advisers continued their projects in their own ways, which were still neither wholly coordinated nor streamlined. Then, in early 2009, Atmar moved on to new responsibilities.

The state of affairs for basic, quality education for Afghan children still raises questions: "Have we seen progress?" "Isn't it well past time for the country's educational leaders and advisers to be better organized?" "Why is it taking so long to see significant change and progress for *all* Afghan children?"

These are, of course, easy questions to pose from a distance of place and time. They are posed, nonetheless, with full acknowledgment of some progress in the Ministry of Education, including Mr. Atmar's noteworthy major steps forward, which, of course, built on his predecessors' work.

The 2007–2012 Strategy had, and continues to have, practical problems. Donor funding for education is still not flowing as effectively or plentifully as aid for public health, rural reconstruction, and development or economic reform. Further, donors and their advisers continue to disagree (or, perhaps, simply to fail to agree—which is quite different, and inexcusable) about how best to support the delivery of basic education services for Afghan children and youth.

Lacking forceful and effective redirection, both innovation and progress continue for good or for ill. Based on years of experience of just getting on with things (which is what Afghan communities, teachers, NGO supporters, and others did in many places throughout the post-communist decades), some communities are doing just that. Yet, inequity (particularly with regard to access) and disparity (particularly with regard to quality) persist, too. Communities with access to educational services supported by NGOs *generally* were accorded a high(er) quality of service and access to a wider variety of lessons as well as better textbooks and other learning materials, while students at government schools are generally less fortunate, particularly in rural villages and towns.[15]

Lest we forget: Afghanistan remains an emergency situation, and a reescalating one at that.[16] By global standards and expectations, all progress toward meeting children's education, protection, and health-care rights is noted—and noteworthy.

Classrooms from Kabul to Kandahar city, from Herat city to Bamyan town, and from Mazar-i Sharif town to Jalalabad center and thousands of smaller towns and villages in between continue to function with varying degrees of effectiveness and excellence or, by turns, mediocrity and inadequacy. The ministry and its persistently varied and sundry advisers and donors also pursue a vision, however out of focus, shortsighted, or improperly illuminated it may be. Western military entities in their various guises continue to involve themselves—some would say inappropriately and, worse, perilously—in school-building and delivery of classroom supplies.[17] But some communities are benefiting from this involvement—verifiably to their peril in certain places.

Yet, education is happening.

So, again, the answer is "Yes" to the question: "Is there progress in education for Afghanistan's children?" Yes, despite the fact that in the years since hope sprung anew is 2002, tens of thousands of Afghan boys have reached puberty without the benefit of even a modestly useful primary education, let alone knowledge and skills to help them become functioning adults. An equal number of Afghan girls have reached the age of eligibility for marriage without the benefit of literacy and numeracy skills, and without essential knowledge of their reproductive biology and rights to help them take on the roles required of them.

The state of education in Afghanistan in mid-2009 is a labyrinth composed of circuitous paths to interesting but inconclusive or inconsequential places, and to some dead ends. In circling around, much ground has been lost—or, more accurately, not gained—in the last seven years. Yet, these very years and prior decades' lessons learned are nothing if not instructive. The certain, expert knowledge, bright ideas, and worthwhile opinions of legions of education experts and advisers cannot be lost now. The confidence and capabilities of Education Ministry staff cannot be discounted or ignored. The answer must continue to be a "Yes!" when asked, "Are things getting better?"— even when we are not fully confident in the affirmative. The cost of "No" and negativity is too high; and, a non-affirmative response is not warranted even when certain facts suggest that things are not progressing well in support of Afghan children's educational rights.

This is not to excuse the Afghan government (and its funders and advisers) for failing to ensure all Afghan children their right to education. For this, they must be called to account. However, given the state of affairs highlighted by the select facts and anecdotes shared above, and by statistics not yet explicitly outlined—for example, 73,000 classrooms are still needed to ensure a seat for each enrolled Afghan child[18]—it is clear that progress is haphazard.

Yet despite the incalculable deficit of female teachers and the criminally inexcusable absence of millions of textbooks from classrooms (books paid for by foreign aid and printed in various foreign countries), one still cannot say, "No, things are not progressing."

Today, children have access to thousands more classrooms than there were seven years ago. Innovative and promising initiatives are increasing the number of well-qualified women teachers, whom Afghanistan needs if girls and young women are ever to make progress toward verifiably better lives and life expectancy. Furthermore, when Minister Atmar moved on to new challenges, another influential and respected minister was named, with the expectation that he will build on past efforts and forcibly seek to rectify shortcomings. Donors continue to support basic education initiatives, which are led, more strongly than in the past, by a cadre of Afghan educationists and more seasoned ministry functionaries. More significantly, communities and local-level education ministry officials—teachers and principals—are continuing efforts to change and to make progress locally. Of course, many communities have been doing this for years, if not decades.

There is progress, just not enough; and it is neither fast enough nor good enough.

There are undeniably horrifying and heartbreaking reports of regression in parts of the country by Taliban and other anti-government, anti-education, anti-female elements. Such reports cannot go unheeded and unaddressed. Yet, reports of educational success in countless Afghan communities—and of outright defiance in some cases where opponents of peace, progress, and prosperity seek control—must be recognized, celebrated, and, most important of all, noted for future reference. Indeed, an endless source of dismay for me throughout all my Afghanistan years was the media's evident imperative to report bad news: setbacks and attacks, failings and shortcomings, horror and pain.[19] All were and are important stories to tell, no doubt. Yet, everyone who takes or is assigned a role in the story of Afghanistan's future knows that a major plot line must, without question, be the story of the ultimate promise and delivered guarantee of education for all Afghan children and youth. The story of the chemistry teacher and her inspired students must be told, as must the decision of one district's leaders in Kabul Province to demand five new girls' schools when, just a few years back, almost all families there had refused to let their daughters attend school past grade three, in part because there were no women to teach them. The story of a threefold increase in girls' enrollment in one Sar-i Pul Province school must be told along with that of a nearby grade-twelve classroom with equal numbers of girls and boys.

In pursuit of further progress the Afghan government, especially the ministries of education and higher education, must collaborate and redouble efforts to proceed with the 2007 Sector Strategy, which must be revisited, affirmed and updated. Both must agree on rational expectations and a deliberate timetable for teacher assessment and credentialing so that most of the 78 percent of teachers who are not properly credentialed and qualified to teach can become so. This must be done in a non-punitive way for the sake of teachers (fully credentialed and otherwise) now in classrooms. It must be done with an eye toward ascertaining more clearly and specifically just what are the levels of knowledge and skills among Afghanistan's teacher cadre. No doubt, suspicions and anecdotal evidence of ineptitude, or worse, will be proven. If there is seriousness and integrity in the endeavor, the chemistry teacher and her competent colleagues will be recognized and praised. Without question, everything in this process—teacher support and credentialing, school management improvement, and so forth—must be done *now* if a promising way forward for Afghanistan's future—that is, today's students—is to be embarked upon.

In starting to establish and verify aptitude and promise in the current teaching force (by whatever nascent tools and processes now available), the Ministry of Education can determine a way forward for (1) continuing education and professional development for teachers (that is, for those assessed eligible to continue teaching as well as able to grow into leadership roles) and (2) new teacher education for those qualified and interested to enter the teaching profession now. On this latter point, a major comprehensive review of the national teacher training college system is warranted to complement and build on past piecemeal reviews by different donors and advisors; and innovations must be demanded. This is especially true if families are to allow or, better yet, encourage their daughters to become teachers. This may be possible if teaching is newly seen as a valuable and noble profession, not unlike being a doctor or an engineer. In fact, one of my strongest opinions at the end of my Afghanistan years is that teaching must become a recognized and valued profession for which teachers are reasonably well paid and praised. Or, better yet, why not view and value teaching in an entirely new way that acknowledges that the nation's way forward requires an educated populace that values education for all?

The government and its advisers—and, in this case, these must be more than just educational advisers[20]—must accelerate attention, including funding and expert advice, to teacher education and upgrading. They also must continue funding textbook development and printing, and assiduously ensure accountability for content, material quality and reasonable cost containment. Most of all, they must guarantee textbook delivery to classrooms—and not to markets where only children whose families have expendable income can afford the schoolbooks that other nations' citizens have paid for in the expectation that all Afghan school children would have them for free.[21]

Also without question, donors must continue to fund school construction, and, again, assiduously ensure accountability for design, quality, and proper use of schools. Education buildings should not become monuments to foreign donors' ideas of what Afghan schoolchildren should have (based on Western and Eastern standards), nor, more scandalously, army outposts or a warlord's horse stable. There is no doubt that building schools is a very serious concern for the government and its donors, not least because so much time and ground has been lost during the past seven years. The construction enterprise, no matter what the product (school, clinic, police station, and so on), is one that lends itself to corruption worldwide, and there is no reason to think that things can be different in Afghanistan. Still, lessons have been learned through the years, and better progress must be made now and into the future.

In addition to the basics (classrooms, texts, qualified teachers), support for Afghan families' efforts to ensure their children are ready to benefit from school is required. The ministries of education and social affairs must continue making progress not only on policy for early childhood development and education but also in experiments with different venues, for instance homes, mosques and schools, and approaches to the sector.[22] As with the example of primary schools, and despite the dearth of qualified teachers, the government plan says that ECD teachers must have bachelor's degrees. Similarly, government initiatives exploring the role of preschools and ECD programs are envisioning school-based, teacher-led kindergartens. The experiences of some NGOs in Afghanistan and elsewhere in South and Southeast Asia with home-based, volunteer-led early childhood programs must be considered for the excellent results they are known to have.

As with early childhood education, it seems that the *au courant* catchphrase about not letting the perfect be the enemy of the good is apt for education in Afghanistan in 2009. Having engaged in the past seven years' efforts to help bring development to Afghanistan, it is clear to me, as it surely is to those wiser and more seasoned than I, that there are many and varied ways to ensure education for all. These many and varied ways must—starting now—be considered systematically and then accepted (or rejected). Accepted approaches must be prioritized, funded, supported, and sustained as planned using the aforementioned healthcare model that is in its seventh year of continuous implementation, verifiable expanding impact, innovation and correction, and further improvement from lessons learned.

Once the Ministry of Education's key components (teachers, classrooms/schools, textbooks/teaching-learning materials) are organized *nationally* (not regionally or provincially, not randomly or haphazardly) and the ministry is committed to build on past sustained successes, mistakes made and lessons learned, the notion, "Yes, more progress can be made in ensuring education for all Afghan children," will gain currency, too.

By way of a coda: Anyone who has followed the downward spiral in security and personal/family safety in Afghanistan will understand that there will be no meaningful progress if the country returns to sustained, widespread unrest. Already, much disruption to education and healthcare service has occurred in Afghanistan's south, southeast, and east and sporadically in the north and west, too. Alarmingly, much of the disruption has been very close to Kabul, which somehow symbolizes heightened peril for its proximity to the country's titular power. These comments and speculations are significant only if the situation in the country does not get much worse in the near term. If things go status quo or, quite possibly, improve, then my optimism remains intact, not least because of the indefatigable spirits and hopes of the Afghan people.

Still, I remain adamant that there is no way for hearts and minds to be won by foreign militaries—not even Provincial Reconstruction Teams.[23] Nor will they be won by any civilian surge (as some are characterizing plans for additional U.S. support) connected, in any way, to any military force, including PRTs. Rather, I am certain there is no better way to win hearts and minds than by supporting, outright, Afghanistan's civil society, which is clearly identified and includes parents, teachers, community leaders, and relevant ministries' staff, to protect, educate, and care for the health of its children. If this can be done in Afghanistan—and I believe it can—there is reason to believe that millions more children affected by emergencies that disrupt their education (sometimes even before it begins) worldwide can have their right to education guaranteed.

17 | Child-Friendly

The Dual Function of Protection and Education in Myanmar

NI NI HTWE AND MAKIBA YAMANO

Cyclone Nargis had a severe impact on children in Myanmar when it struck in 2008. It changed the lives of children who lost their parents, siblings, and friends, children who found themselves in unfamiliar locations and children who took on additional responsibilities to rebuild their lives.

In the aftermath of an emergency, education, especially when it becomes the primary basis of child protection, is every bit as important as food, water, shelter, and health care. Child protection is an essential part of disaster response because without it children become vulnerable to disease, abuse, accidental separation from their families, and even human trafficking.

This chapter emphasizes the importance of education in emergencies in relation to physical and psychosocial protection of children. World Vision's (WV) experience through the emergency and rehabilitation phases of Cyclone Nargis has reaffirmed that education enables children to regain a sense of normality, and is one of the strongest forms of protection that communities, aid agencies, and the government can offer in a disaster.

Background of Cyclone Nargis in Myanmar

Cyclone Nargis struck Myanmar on May 2, 2008, causing widespread destruction and devastation in Yangon and the Ayeyawardy Division.

The official death toll stands at 84,537, with 53,836 people still missing, and 19,359 injured. Some 2.4 million people were severely affected by the cyclone, out of an estimated 7.35 million people living in the affected townships.[1]

Child deaths are believed to have been substantial, although fatalities disaggregated by age are not available. Many families were separated, which put unaccompanied children at risk of abuse, exploitation, and lack of access to humanitarian assistance.

Education facilities were seriously affected by Cyclone Nargis, with approximately 60 percent of schools damaged or destroyed[2] and widespread loss of school furniture and teaching/learning materials. An increase in the school dropout rate was expected, due to the loss of earners in the family.

Coordinated Response to Child Protection

Immediately after the cyclone hit, the UN set up eleven clusters to coordinate the various sectors of work that agencies were preparing to implement. The needs of children were primarily cared for within two clusters: the Education cluster,[3] which took the lead on building and repairing infrastructure (such as schools used for formal school education), and the Protection cluster, which was created for the Protection of Children and Women (PCW) and led nonformal education activities, including awareness-raising of safety and survival skills, and psychosocial assistance for children. In the case of Cyclone Nargis Response, this PCW cluster enabled humanitarian agencies to agree on common priorities and approaches, as well as to coordinate with government counterparts. The close coordination effort among agencies resulted in increased protection for the affected children of Myanmar.

The cluster members initially consisted of child-focused agencies with long working histories in Myanmar, including UNICEF, EMDH, Save the Children Alliance, World Vision, and national NGOs. Capitalizing on the existing networks and experiences among agencies, the

PCW cluster was able to draft a common plan of action for Cyclone Nargis Response in close coordination with the Ministry of Social Welfare.[4] Within one month, a draft of the National Plan of Action (NPA) for Child Protection in Emergencies was created,[5] based on the UN Convention of the Rights of the Child and Myanmar Child Law.[6]

Defining its goal as to "fulfill the survival, development and protection needs of affected and vulnerable children, including those separated from their families, unaccompanied and orphaned by Cyclone Nargis," the NPA provided the following six specific objectives for all the stakeholders working for the affected children:

- To promote family unity and where possible prevent the separation of family members.
- To reunify separated children with their parents or extended family.
- To ensure adequate care and support for separated, orphaned, and other vulnerable children and their families.
- To promote community-based psychosocial well-being and the establishment of safe environments for children and women.
- To prevent violence, abuse, exploitation, and neglect.
- To work with government, partners, communities, and families on key protection and rehabilitation/reintegration interventions, including identification, documentation, tracing and reunification, establishment of appropriate forms of alternative care, services to prevent family breakdown, services preventing and responding to sexual abuse, violence, and exploitation.

In order to achieve these objectives, one of the common approaches agreed among the cluster members was the establishment of safe spaces for children, where they could learn and play. The PCW cluster members established 380 safe spaces serving more than 65,000 children.[7] The following section will focus more on the approach called "child-friendly spaces," looking at its dual value for protection and education.

Child-Friendly Spaces

The term "Child-Friendly Space" (CFS) refers to a space set up specifically for children in crisis, usually after a disaster, emergency, or conflict. Its purpose is to address both physical and psychosocial needs of children in a stable environment that invites trust. A CFS can be located in a school compound, a community center, a tent, or an open space in a camp or community; however, the one consistent rule is that it must be a place where the children and community feel safe. A CFS is a structured and safe place where children and youth meet other children to play, learn competencies to deal with the risks they face, be involved in some educational activities, and relax. It gives the children the sense of safety, structure, and continuity that provides support amid overwhelming experiences. In this sense, CFS has an interconnected component of both education and protection through education for survival skills.

During the emergency response, WV's child protection team established 117 CFS serving the needs of nearly 17,000 children[8] across the Ayeyawardy and Yangon divisions. In the meantime, other emergency response activities were taking place that enhanced the welfare of children such as food distribution, temporary shelter, water, sanitation and hygiene, health services, and emergency livelihood activities.

WV's first CFS opened in Yangon in the third week of May. More than 300 children attended the small CFS on a daily basis in some communities. Volunteers needed to make schedule shifts to allow every child to participate at different times.

In the initial few weeks after the cyclone, a Buddhist monastery compound was used as a temporary CFS location. Both children and community members suggested the monastery as a physically and spiritually safe place. However, more spacious and safe CFS locations were needed to accommodate more than seventy to eighty children at a time for a few months. With children's and the community's help, proper CFS were established with bamboo and tarpaulin on higher ground that protected children from floods.

Activities in the CFS were designed to be inclusive, allowing diverse groups of boys and girls from different ages and areas to attend. Detailed plans for one to two weeks of activity were prepared in advance, in order to provide structured learning. Various activities were designed and prepared for different age groups and genders. Traditional children's games and songs were selected by a group of experienced WV national staff, and a guidebook for games and a songbook were created.

CFS volunteers were recruited from the local community in order to promote sustainability. Volunteers were expected to become resources to the community, sharing the knowledge and skills of child protection after the program had been completed. Experience working with children (such as as teachers or youth leaders) was preferred, but this was not always possible. In order to offer quality care and support, WV provided recurrent trainings on different topics related to child rights, child protection, childcare, psychosocial response, health and nutrition, and other child-focused topics. Gender balance was also an important consideration when choosing volunteers. In order to prevent and minimize the incidence of gender-based abuse and exploitation, at least 50 percent of the volunteers were female.

Recruited volunteers and staff:

- Designed child-friendly activities appropriate for different ages and genders and for children with disabilities.
- Created a warm, welcoming environment where children felt safe and relaxed.
- Educated children with important information about the new situation around them.
- Identified other vulnerable children not attending the CFS, who were orphaned, injured, abused, disabled, or had missing family members.
- Referred children and their families to other agencies who could provide necessary assistance.
- Advocated to the local government when necessary.
- Conducted parental training and awareness training for the community on child protection and child rights.

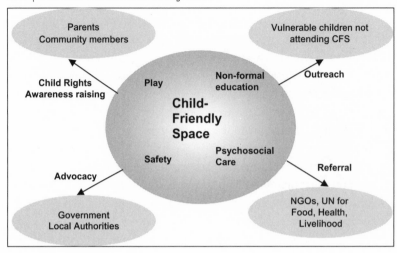

Figure 17.1. Activities around CFS.

Five Primary Functions of CFS

Securing Safe Places for Children

With schools and houses destroyed by Cyclone Nargis, children did not have adequate safe space to meet and play because their existing protective environment had been destroyed. With adults busy rebuilding their livelihoods, there was a decrease in the amount of supervision provided to children in the community. A lack of supervision could expose children to different risks including sexual or physical abuse, trafficking, separation from parents, injury, or even death. Some families were displaced from their home village to transitional shelters,[9] where the risk for children could be even greater.

CFS were located in the safe spaces, and offered supervision of children by trained adults. One volunteer was assigned to every twenty-five to thirty children to facilitate activities. This arrangement prevented strangers approaching children without the notice of volunteers. Each CFS kept a logbook in order to track visitors.

Psychosocial Assistance for Children

Many children affected by the impact of the cyclone experienced minor psychosocial distress during the initial months, including fears of strong winds and heavy rain, nightmares, or depression. Parents were often surprised at these signs of distress; however, signs of stress are a normal reaction to such events and generally do not require specialized psychiatric assistance. CFS are a safe space for children to play, meet with other peers, and express their feelings through drawings, songs, and role-playing. By participating in these activities, children released their stresses and switched off from their worries and concerns. In the case of a child presenting signs of significant persistent psychosocial distress or trauma, the child would be referred to a specialized agency.[10]

Teaching Survival and Protection Skills to Children

Children faced many changes after the cyclone, and teaching them basic life and protection skills was important for their survival. With limited sanitation facilities and lack of access to clean water, poor hygiene practice could result in increased morbidity and mortality. Using easy language and visual aids, children learned basic health and hygiene practices.

In transitional shelters children were sometimes faced with threats of abuse, and children without caregivers faced risks of trafficking. In the CFS children learned how to protect themselves from these risks. A coloring cartoon booklet was distributed to each child, showing them the types of abuse they could face, how to avoid them, and how to respond if they became a target of such an act.

Children also learned about what a cyclone is and how to be better prepared next time.

Smooth Transition to Formal Education

The cyclone hit the country in the midst of the school holiday season, and so children did not have safe places to meet, play, and learn. The

CFS worked as a transitional social educational space for children, enabling them to resume their daily routines by attending activities and making new friends. This routine assisted with the transition into formal education when schools reopened. Some of the children who had dropped out of school before Cyclone Nargis went back to school with their new friends, stimulated by the friendship and learning opportunities developed in the CFS.

Providing Parents Time to Rebuild Their Lives

Providing safe and supervised spaces for children gave parents time to collect humanitarian assistance, rebuild their houses, and seek livelihood opportunities. While working far away from home, parents did not need to worry about the safety of their children if their children were enrolled in CFS activities regularly.

Major Challenges for CFS

Approaching the Most Vulnerable Children

The number of children participating in the CFS was greater than expected. Nonetheless, many vulnerable children who needed the most assistance were unable to participate in the program for a number of reasons:

- Many children with severe disabilities were not able to participate because of the lack of regular transport assistance.
- A significant number of children had to work to assist their parents in livelihood activities and could not participate because of their workload.
- Out-of-school children (children who had previously dropped out of school) faced some discrimination in the community, which made them feel ashamed to participate in CFS. WV received several "complaints" from caregivers of schoolchildren claiming they did not want to see out-of-school children in the CFS, as their children might catch bad behavior and practices from them.

Continued effort for outreach and family visits enhanced the participation of vulnerable children to some extent. However, it was a constant challenge to cover the different categories of vulnerable children in one CFS.

Consideration of the Timing to Close CFS

When schools began to reopen in June 2008, staff held discussions with children and communities about whether they would like to continue CFS activities. The initial intent of the CFS was to bridge the gap until children were able to reengage with formal education. During the discussions, both children and communities strongly recommended that the CFS remain functioning in addition to the primary school, highlighting that:

- Children do not have enough playtime in schools.
- Children learn different knowledge and skills at CFS, including life skills.
- Many schools are open in the morning or evening only, and children still need to have a safe place to play when they are not in school.

During the transitional phase from emergency response to livelihood recovery, WV began the handover of CFS management to the communities. The majority of the community members acknowledged the importance of the CFS and the ongoing needs for these spaces. They did, however, express concerns related to sustaining volunteers and financial resources. Despite the high contribution of labor and building materials for CFS from the community, they worried about sustaining the expenses for both the volunteers and a replenishment of the playing materials.

WV discussed these issues with community members making the suggestions that:

- The CFS should open less but at regular times (such as once a week).

- Volunteers should be paid with commodities (rice, oil, etc).
- Community funds should be pooled to maintain the spaces and playing materials.

Yet in reality, continuing the management of CFS remains a challenge for communities still struggling to rebuild their livelihoods.

Key Lessons Learned

Training and Supervision of Volunteers Improves CFS Quality

In the aftermath of the cyclone, the urgency of setting up the CFS meant that extensive and lengthy training for the newly recruited staff and volunteers was not possible. Instead, a basic child rights, child protection, and CFS management training was conducted within one week, and follow-up training and coordination meetings were conducted weekly. Teaching materials were provided on a weekly basis and volunteers seemed to handle the educational component very well with this approach.

With a mass recruitment of new volunteers, guidance and training were critical in ensuring a nurturing, safe, and friendly environment for children. Experienced child-protection staff members were tasked to monitor and supervise a CFS at least two to three times a week, staying in each space for several hours to observe how volunteers communicated with children. This regular monitoring mechanism maintained the quality of the day-to-day program and rapidly enhanced the capacity of volunteers.

Importance of Education for Parents and Communities

Child protection cannot be achieved by targeting children only. Their caregivers and community members should be involved in activities, to build the sense of responsibility about their children's protection. Child-protection training for caregivers and community members should be one of the care activities to build a protective environment around children.

Psychosocial awareness-raising for caregivers also improved the relationship between children and parents, enhancing family unity. Some parents did not understand why their children behaved differently after the cyclone—becoming emotional, irritable, seeking attention, or being afraid when their parents were out of eyesight. Training for parents assisted them with ways to respond to and communicate with their children in such cases, and how to manage their own emotions and psychosocial stresses. It improved the resilience of families to cope with difficult circumstances in situations such as the post-Nargis period presented.

Multisectoral Assistance Is Effective

Vulnerable children, such as separated children or children cared for by the elderly or persons with a disabilities, can often miss out on humanitarian assistance, for they face particular difficulties in participating in livelihood activities such as food for work programs due to their limited physical strength. This has resulted in the low well-being of dependent children.

Separated children were often temporarily cared for by relatives as an additional family member, which can result in smaller rations for the caretaker's family. In the CFS, volunteers and staff kept records of especially vulnerable children, including such cases. The list also covered the children who could not participate in CFS through outreach work.

During the emergency phase, the list of these vulnerable children was shared with the food and livelihood response teams in order to prioritize them in the beneficiary selection. This encouraged families to keep the children under their care and help them pay the cost of food, education, and healthcare for these vulnerable children.

Changing Practices for a Better Tomorrow

Behavioral Change: A Sign of Lasting Confidence

CFS had an overwhelmingly positive impact on children and caregivers. Feedback sessions from children and parents pointed out that "behavior change" was the most visible impact on their children. Grief, extreme distress, a sense of loss and hopelessness were all commonly felt emotions after the cyclone. The CFS helped children work through these feelings. WV's external program evaluation report by TANGO International noted that the establishment of CFS was "among the most effective components of the Cyclone Response Program." It further noted, "Parents claim that participation in CFS has greatly eased the psychological stress of children in the months since Cyclone Nargis, has ensured their continued educational development amid a difficult period, and enhanced their self-confidence."[11]

In some cases, the attitude and behavior of children with disabilities improved over the period before the cyclone, which surprised parents and made them realize the potential of their children's learning ability.

When she started joining CFS, she was silent and could not communicate with others. She could not trust anyone and held her belongings tight in her hands. She did not communicate with volunteers or her friends. Now her attitude and behavior changed and improved, after she realized that CFS is a safe place to play with her friends and to enjoy songs and games. (A father, Haingyi Township)

[After the cyclone] our child was aggressive and did not listen to us well. Now she looks happier to learn, speak politely, and keep her belongings neatly. We are happy and pleased for our children's improvement. We don't need to worry for our children as they are staying in this center safely. (A mother, Haingyi Township)

Feedback from children participating in CFS was also very positive. Children indicated that they were learning to acknowledge their negative emotions and turn them into positive ones. Here are some comments children made:

Before Nargis we were only reading school lessons at home and were bored. When Nargis hit our village, we got sad. But when the CFS opened, our emotions changed and we felt happy again. (A girl, Bogaley Township)

I am very happy to go to CFS because it is fun to meet with friends. I learn how to communicate politely to others and get knowledge about not to be a naughty boy. (A boy, Haingyi Township)

In CFS, I can learn about health and child rights, and get snacks. (A girl, Dedaye Township.

At the CFS, teachers don't beat us, so we are happy. (A boy, Bogaley Township)

Further evidence supporting the high impact of the Child Friendly Space was noted in continued requests from neighboring villages to establish CFS in their areas. Village leaders and schoolteachers cited "behavioral change" as the biggest reason why they would like to set up CFS even at their own expense.

I met children from the neighboring village, and they all seem happy and very polite. I heard that these children were attending a child-friendly space, and I believe it can bring positive change in our children too. (Village schoolteacher, Pyapon Township)

Feedback from parents and communities confirmed the positive psychosocial impact and the development of greater self-confidence among attending children. Such behavioral changes could not be achieved by only providing "life-saving" assistance such as food and water; the psychosocial needs of children must also be acknowledged. With mental well-being support structures in place and assurance of a safe environment, children are likely to recover from their psychological distress a lot quicker. WV's work with children in the disaster area helped rebuild a sense of self-confidence and instill a belief in a positive outlook for the future. With newfound self-confidence, children are better able to say no to abusers who try to take advantage of them, and to discuss such incidences with trusted adults.

Improving Knowledge and Practice: Educating Caregivers and Community

The child-protection work in the emergency phase of our response focused on awareness-raising and educational training for children. The next step was to plan interventions that educated caregivers and community members about their roles and responsibility to protect children within their family and community. The training included orientations for Child Rights and Child Protection, discussing the negative impact of corporal punishment at home or school and how to replace this approach with positive discipline, or the risk of sending children to orphanages and what difference it would make if children stayed in a family environment.

The external evaluation report suggests a high awareness of norms around child rights among adults in the target area; at least 96 percent of households agreed with all four child-rights statements:

- Children have the right to be protected from abuse and neglect.
- It is against the law to sell children to another person.
  Vocational training and nonformal education should be provided to out of school and disabled children.

Television, radio, and newspapers should only present information that will assist children.Evaluation results also suggest that awareness of child-protection issues is high, given that over 93 percent of all respondents agreed, or strongly agreed, with four of the five statements.[12]

In addition to the awareness-raising sessions, community-based Child Protection Committees were formed to work as "watch groups." Members of the committee were expected to take leading roles in protecting children in their own communities by raising awareness about child rights and child protection, using leaflets and visual aids provided by WV, identifying the vulnerable children in their community, referring vulnerable children to appropriate government or humanitarian agencies to meet their immediate needs, and facilitating care

Table 17.1. Agreement on Statements Related to Child Protection, by Division and Percentage

	YANGON	AYEYARWADY	TOTAL
It is wrong to punish a child by hitting him or her			
Strongly agree	17.4	29.0	24.6
Agree	42.9	39.7	40.9
Neutral	30.8	24.4	26.8
Disagree	7.4	5.0	5.9
Strongly disagree	1.6	1.9	1.8
Protecting children is a communal responsibility			
Strongly agree	39.5	48.1	44.8
Agree	57.4	49.7	52.6
Neutral	1.3	0.3	0.7
Disagree	0.0	0.0	0.0
Strongly disagree	1.8	1.9	1.9
Adults should listen to and involve children about topics that concern them			
Strongly agree	26.8	42.7	36.7
Agree	61.3	53.5	56.5
Neutral	8.2	1.0	3.7
Disagree	0.0	0.2	0.1
Strongly disagree	3.7	2.6	3.0
Children should be given free time to play daily			
Strongly agree	28.9	48.1	40.8
Agree	62.9	49.4	54.5
Neutral	5.8	0.8	2.7
Disagree	0.0	0.0	0.0
Strongly disagree	2.4	1.8	2.0
Children should not miss school for work			
Strongly agree	42.4	61.1	54.0
Agree	48.9	35.8	40.8
Neutral	3.7	1.6	2.4
Disagree	3.2	0.0	1.2
Strongly disagree	1.8	1.5	1.6

and response in cases of extreme abuse. The committees were expected to be linked with the government's Child Protection systems, such as the township Convention of the Rights of the Child committee, responsible for coordinating the prevention and response to child-abuse cases.

One example of improved community understanding of child protection came from the Bogaley Township, where strangers entered a community with a Child Protection Committee, disguised as journalists. After the town discovered that they were not journalists, suspicion was raised and an alert was shared between all the 66 Child Protection Committees in Ayeyawardy Division. Upon receiving this "alert," some community members took turns watching over the CFS in order to prevent human trafficking or abuse.

These coordinated efforts within the community for child protection had not been observed prior to the Cyclone Nargis Response. Community members also developed confidence in these successful prevention measures and gained experiences for creating protective environments for their future generations.

Conclusion

Cyclone Nargis undoubtedly left a severe impact on children. Establishing CFS as part of the humanitarian response ensured that children had a safe place to play and learn through nonformal education, allowing them to regain a sense of normality and receive assistance in preparing to deal with their new environment.

Child-protection awareness to caregivers and community members improved the protective environment for the children. By involving trusted adults such as parents in child protection activities, children were also able to develop psychosocial stability and family unity.

Evaluation from both children and the community confirmed that CFS activities helped switch off children's minds from their grief and regain the confidence to start their new lives even under difficult circumstances. When picking up the pieces after a disaster, humanitarian aid workers often realize the potential to build better futures for children. In Myanmar, this meant not only the provision of water, food, and shelter, but also providing hope and strength within children and communities to stand on their own feet.

The importance of playing and learning for children in emergencies should not be underestimated. Child-protection programs, including setting up CFS, should be an integral part of the life-saving interventions during a humanitarian response, to enhance the communities' potential to rebuild better.

18

After the Storm

Minority School Development in New Orleans

JUAN RANGEL

In 2005, when natural disaster hit New Orleans, Louisiana, in the form of Hurricane Katrina, life came to a standstill. No aspect of city life was left untouched by the hurricane. It destroyed not only neighborhoods but also the city's infrastructure. In addition, it completely uprooted the public education system of Orleans Parish. The public schools in Orleans Parish before the hurricane were far from delivering an adequate education to their students. Thus, this chapter is a story about constructing an entirely new and better system of public education to meet the needs of families in New Orleans, and, in particular, about the contributions of the United Neighborhood Organization (UNO) in developing functional programs for newly arrived Hispanic workers and their families. Natural disasters almost always result in significant demographic changes. New Orleans post-Katrina was no exception. However, creative response to chaos can produce lasting beneficial responses, and this has been the case of new charter schools servicing a minority population in New Orleans.

It is difficult to overstate the devastation Hurricane Katrina wrought on the city of New Orleans. More than 1,500 residents perished in the storms and during the evacuations. In addition, 80 percent of the city was covered by an average of seven feet (2.1 meters) of water. After several weeks, when the floodwaters receded, much of the infrastructure in areas of the city, especially the Lower Ninth Ward, was destroyed and unusable. Houses, trees, mobile home parks, and other structures were literally lifted off the ground by the power of the winds and the strength of the water, which broke through the levee. In the worst affected areas of the city, what had not been washed away in the storms was lost to an epidemic of rot and mold emerging from the infamous swamp heat of New Orleans. The floods and the mold consumed homes, offices, businesses, and schools, not to mention possessions, inventories, supplies, and records.

The destruction so thoroughly uprooted the lives of the inhabitants that they had no choice but to begin anew elsewhere. People were evacuated as far north as Canada, and to Houston, Atlanta, and Chicago. In fact, the whole of the United States welcomed refugees of the hurricane. Within a year of the hurricane's damage, the city's population was only half of what it had been before the storm. Those who did return encountered chaos, from rampant crime to relocation in poorly planned and unorganized mobile home parks.

Trying to pick up the pieces in the aftermath presented unimaginable difficulties. How, for example, does one file an insurance claim on a damaged house when all records of deeds and titles have been lost? The rebuilding effort that followed the storms would encounter similar difficult predicaments. It was clear that the life of the city had been so irrevocably changed by the storms that "reconstruction" meant starting over again from scratch. The need for cleanup, rebuilding, and reconstruction after the hurricane attracted thousands of Hispanic workers, because the previous New Orleans workforce had been dispersed. This new population of workers and their families created new challenges such as developing housing and social services as well as educational services.

In the midst of these new realities brought by the hurricane, restoring public education was a focal point. The vast destruction and new challenges facing New Orleans led to a reinvention of public schools that required the help of myriad outside actors to provide their expertise and services. A new school system, despite the long odds and significant risks, would have to be created.

UNO, a civic organization based in Chicago, has a successful history of organizing Hispanic neighborhoods as well as running charter schools primarily for this population. UNO volunteered to assist in the task of rebuilding the public schools of post-Katrina New Orleans.

This backdrop sets the stage for the challenges that confronted the post-Katrina public school system. As with the rest of the city, the storms rendered most public school facilities and administrative offices unusable. In addition, many other capital outlays such as school buses had been wiped out, as had essential school supplies from textbooks to desks, all of which would need to be replaced. Beyond these material needs, many school personnel—teachers and administrators—left the city for good in the wake of the storms, creating major staffing concerns.

These challenges created major obstacles in reopening schools. However, they hardly represented the sole extent of the problems confronting the New Orleans school system. In fact, before the hurricane, New Orleans had earned a reputation as one of the worst school districts in the United States, with major corruption and some of the lowest student test scores in the nation. Sixty three percent of the schools in the Orleans Parish were deemed by the state of Louisiana to be "academically unacceptable" based on 2004–2005 state test score results.[1]

The situation was so dire that Louisiana voters approved a referendum for an amendment to the state constitution allowing the state to take over the worst of the failing schools under the auspices of a "Recovery" School District. This district answered directly to the Louisiana Department of Education's Board of Elementary and Secondary Education (BESE) rather than to the local Orleans Parish School Board. Once placed within the "Recovery" District, BESE had broad authority to provide independent charter operators a chance to staff and run these failing schools.[2] Before Hurricane Katrina, five of the city's public schools had already been taken over by the state under this law and handed over to the Recovery District.[3]

Given the failings of the Orleans Parish School District, simply rebuilding and restaffing destroyed schools would represent a lost opportunity to vastly improve the academic outlook for returning students. Reconstituting the old, broken system hardly represented a humane option for city residents. Through its vast destruction, upheaval, and displacement, Hurricane Katrina left behind a blank canvas that demanded an entirely different educational system to better meet the needs of city residents.

The trauma that Katrina created in the lives of students and their families required deliberate, new approaches to education. Before the hurricane, the city's school system was dealing significantly with issues of unemployment and wrenching poverty. Katrina threatened to make a bad situation worse. Much of the student body of Orleans Parish pre-Katrina came from backgrounds of heavy poverty with all the corollary problems of low attendance rates, unmet physical needs, violence, and general disorder that too frequently typify urban schools. Much of this could be traced back to the status of New Orleans as a city in decline, steadily losing population for nearly twenty years and failing to attract major industry to build on its status as a major tourist destination. Since tourism, hospitality, and its small port remained the backbone of the area's economy, with higher wage industries never materializing, the city's population stagnated, with many residents leaving. Census data show a population peak in Orleans Parish during the early 1980s, followed by a slow decline that would continue uninterrupted until Hurricane Katrina hit.

The aftermath of Katrina threatened to bring even more distractions to the education system. Families no longer had their homes or jobs, while many basic goods and services were even harder to come by. Some families were also split apart, with students returning to New Orleans in order to finish their education alone and unsupervised or with just one parent. The circumstances that students—and therefore their schools—faced were far more difficult than before the hurricane.

There was one other development with regard to post-Katrina life in the city that was of particular interest to UNO: the influx of Hispanic families into New Orleans, arriving to take jobs in the reconstruction industry. Prior to this wave of Hispanic settlement, New Orleans had historically been home to only a nominal Hispanic population. The Hispanics living in the city and surrounding areas were few and largely associated with the shipping industry, due to the active ports handling traffic from Central and South America. As a result, New Orleans had little experience with the needs of low-skilled Hispanic immigrant workers from Mexico arriving to fill jobs in construction and in the hospitality industry, jobs left vacant after the exodus of local longtime residents.

In the midst of the reconstruction efforts, opening schools, along with parks, was a top priority in the interest of giving children diversions. It was deemed important to allow children to return to some level of normalcy as the rest of the city dealt with the trauma of the hurricane. Opening schools also allowed parents some much needed respite from childcare duties while they too tried to recreate stable family lives. However, schools did not begin to reopen until January 2006, four months after the storms, and only then just three charter schools under RSD were able to open their doors at that time. In April, three more RSD-run schools opened to complement the charters, expanding enrollment in Orleans Parish to almost eleven thousand students. However, reports at the time indicated that not all students were able to enroll and that, even among the schools, which opened, severe teacher shortages, persisted. Therefore, many children who would be returning to RSD public schools and charters did not receive much of an education for the entire 2005–2006 school year.

It was not until the following school year, beginning in fall 2006, that the schools were ready to receive students. At this time, just fifty-three schools were open: five public schools run by the Orleans Parish School District, thirty-one charter schools overseen by RSD, and seventeen public schools run directly by RSD.[4] Still, in many ways, the schools and RSD were not prepared. First, even though a year had passed, much of the city's basic infrastructure was still a shambles and essential services such as water and gas were unavailable in certain areas. Second, the strains of the yearlong instability brought on by Katrina began to reveal themselves for students in ways that were difficult to handle. Among those strains were the psychological effects and traumas students witnessed during the hurricane as well as the ensuing disruptions in their lives as they and their families coped with the destruction. This meant, among other things, dealing with the yearlong gap in schooling because of the shortage of seats described above or because of temporary relocations. For kids who were already several grades behind according to test score data, the year off from school was damaging, to say the least. The enrollment figures in the fall of 2006 were only about one-third what they had been during the earlier half of the decade—25,651 students for all public schools in

Orleans Parish. This was a major indicator of the continuing trouble and upheaval with which the city was dealing.

Despite the setbacks, the 2006 school year proceeded apace. However, the number of charter operators was still much less than expected. It was certainly no easy task to invite non-profit charter operators, particularly those from outside the area and with more experience, into the situation that was New Orleans at the time. More than a few major charter operators refused overtures from RSD and Louisiana education officials. For these groups, the lack of city services and general disorder still in the city led them to believe that the probabilities for successfully turning schools around would be low. Even taking on the challenge would unwisely put their reputations on the line.

Between fall 2006 and fall 2007, several major changes began to take shape. Most importantly, in February 2007, a new state superintendent of schools took over Louisiana's Department of Education. He brought with him a sense of urgency to realize some of the reforms that the state had already put forward and to take full advantage of the vast new powers the Department and the RSD wielded to shape the direction of schools in Orleans Parish. As a sign of the direction in which he would take the district, he hired a superintendent of the RSD who was a nationally known education reformer, a former big-city superintendent who had headed both Philadelphia's and Chicago's public school systems.

This confluence of forces eased the path for outside charter groups to open in New Orleans allowing interested operators to prove their educational models. If charter schools could count on the appropriate support from the district, the state, and other nonprofits to ease some of the political battles that have tended to accompany charter expansion, they might have a better chance to focus on education and succeed even amid the extreme circumstances wrought by the hurricane.

Those same chaotic circumstances created by the hurricane still meant that the chances for failure would be high. But charter operators also realized that if they could thrive in a disaster area and transform a complacent education system, they might be able to transform academic models and prospects of poor minority students nationwide.

The wide political support they would have from state and local educational officials would provide them with enhanced opportunities for success.

At the same time, the drastic changes to the status quo caused by so many charters opening in New Orleans and justifying the move with promises of vastly improved results also meant that charter operators would have to perform to greatly heightened expectations. Failing to meet those expectations could be devastating to the reputations of both the charter operators and the national education reform movement in general. The risks were enormously high, but so were the rewards.

Many charter operators willingly stepped into the breach to not only return schools to New Orleans, but instill new hope to students who had been ill served under the system prior to Katrina. They looked forward to the challenge and the opportunity to prove what was possible in schools primarily serving minority students. When the opportunity to apply for a charter arose, UNO opened the Esperanza Charter School in the fall of 2007, its first charter campus outside the city of Chicago.

Our decision to enter the fray in New Orleans did not happen in a vacuum: the charter application was the direct result of massive campaigns by the Louisiana Department of Education and the RSD to recruit outside operators into the city. There was a need to bring in organizations from across the country that already had experience and sizable operations, including the staff resources to begin programs immediately. In fact, RSD had hired the executive director of the National Association of Charter School Authorizers (NACSA) to help use his contacts with lobby groups and charter providers nationally to open schools in New Orleans.

Since the public school system, until that point, had been nearly all black, there was little local experience working with Hispanic students or their families. Since many of the Hispanics in this influx were recent immigrants to the United States, and not just to New Orleans, it represented a particular challenge to help them adjust and become incorporated into American life. This was no easy task in a city that was profoundly torn apart; and it is difficult to imagine a more chaotic situation for new immigrants than that of a decimated city.

Indeed, when UNO's leadership first visited New Orleans in November 2006, the scene was eye-opening. It was only two months into the first full school year post-Katrina, and there was little order. Many Hispanic immigrants and their families were living in motel rooms, vans, or cars. Officials reported the difficulty of locating Mexican families and getting their children into school. They needed help doing outreach to register children of new resident families into schools; they simply did not have the staff capacity or expertise to do so.

New Orleans presented a ripe challenge. The work needed there among the city's newly arrived immigrant population was an obvious parallel to the work UNO was already doing in the Hispanic neighborhoods of Chicago, but it was a far more challenging task. Given the haphazard settlement of Latino reconstruction workers, there were no definable neighborhoods to speak of in a traditional sense. Even locating students to register for school enrollment, as UNO quickly discovered, meant canvassing parks, hotels, construction sites, and even taco trucks, rather than the door-to-door effort the organization was more familiar with in Chicago.

Our work in New Orleans, however, was assisted by a unique history and perspective on organizing and education that is not common to most of our peer organizations. UNO's unique perspective on schools views them as anchors of communities, institutions where immigrant assimilation plays out, and children and families are socialized to the norms of American society. UNO schools are seen as intentional platforms to speed the assimilation of the families it serves through education.

Institutions such as settlement houses or neighborhood clubs and associations that had a strong presence in urban immigrant enclaves at the beginning of the twentieth century no longer exist for the current wave of Hispanic immigrants. However, schools still have the same potential as those older institutions. They have the latent capacity to engage both students *and* their families. It is not uncommon in middle-class suburban schools to see parents or administrators initiate bake sales, start Boy Scout or Girl Scout troops, organize book fairs and reading nights, neighborhood fairs, and other similar family activities. It is rare to see these activities happening at the public schools in less affluent communities.

Schools and, more precisely, education are a key to Hispanic empowerment. Throughout much of its history, UNO had been an active voice in Chicago's education-reform movements: academic success was not just a hope of the organization but an expectation. We sought to prove the potential among low-income Hispanic students for academic excellence. Too often in education circles, poor academic results among immigrant students are written off as the product of their disadvantaged backgrounds or their lack of familiarity with the English language. Those excuses should not be allowed to absolve schools of their responsibility for poor results. In New Orleans, immediate attention was given to replicating our model in this new environment: high academic expectations, strong discipline, and engaging programming for students, parents, and the surrounding neighborhoods. Schools were transformed into community anchors for assimilation. Modifications to the core model were critical due to the particular needs of a population recovering from a natural emergency. Katrina affected the educational system in New Orleans on several fronts: infrastructure deficits, lack of materials and resources, teacher shortages, emotional and physical stress that comes with surviving the aftermath of a devastating hurricane, and displacement and rapid shifts in demographics.

The success of the city's education reforms may have a lot to say in this matter. If the new public school system succeeds in producing a well-educated workforce, New Orleans just might have a chance at coming through this catastrophe in better shape than ever.

19 | The Sudan: Education, Culture, and Negotiations

FRANCIS M. DENG

The word "education" normally connotes schools, formal institutions of learning, for children and youth. This educational system poses special challenges in emergency situations. Education in the broader sense of information gathering and generation of knowledge about a given situation or issue, however, has other dimensions or parameters that are not any less challenging in emergency situations. These include: understanding the dilemmas that emergency situations present for those involved in humanitarian operations, both foreign aid workers and recipient national populations; developing knowledge and appreciation for the cultural context and the values of the people who are being assisted; learning about the root causes of the conflict that has generated the emergency and that are often directly or indirectly identity-related; searching for models of constitutionalism and governance for the management that contribute to the resolution of identity conflicts; establishing mechanisms for peace, security, and stability that are structurally preventive of future crises; and identifying culturally oriented principles for preventing, managing, and resolving conflicts. In addressing these issues, I build on personal experiences and other previous studies that I have conducted in related fields.

Formal Education for Children and Youth in Emergency Situations

It is widely recognized that education is central to the development of a person's sense of identity and dignity.[1] And yet, in many developing countries, especially in Africa, only a fraction of children of school age have access to formal education. Even in this category, competition for higher education is intense and only an even smaller percentage manages to rise up the educational ladder. In some cases, girls do not enjoy equal access to the educational opportunities available for boys.

The result of the disparity in education is that even within one family, some rise to higher levels of education, which open doors to employment opportunities and professional careers, while others remain relatively incapacitated by lack of functional skills and knowledge in the modern context. A class structure created by education or lack of it emerges to divide even members of one family.

Emergencies are usually created by armed conflicts, which generate massive displacement, a problem that affects some 30 million people in more than fifty countries and with which I was engaged as Representative of the Secretary-General on Internally Displaced Persons (IDPs) for twelve years (1992–2004). In emergency situations, most displaced children have no access to formal education, sometimes condemning a generation in situations of protracted conflicts to dismal prospects for the future. Even in urban centers, where services are often better than in the rural areas, the children of those who are forced to flee to towns, often living in shanty peripheries of the urban centers, have little or no access to education.[2]

Even when educational facilities are available in the areas of displacement, parents may be so destitute that they cannot afford the school fees, uniform or clothing. As Representative of the Secretary-General on IDPs, I came across situations where children were not attending school for reasons that were as basic as lack of shoes. In some situations, where people are displaced in areas where violence remains rampant, parents may find it too dangerous for their children to travel to school.[3]

The right to receive at least basic education for children is provided for in international human rights and humanitarian law, even in times of tensions, disturbances or conflict. This right is guaranteed by Article 13 of the Convention of Economic, Social and Cultural Rights (CESCR) and Articles 28 and 29 of the Convention of the Rights of the Child (CRC) as well as by regional instruments. Although Common Article 3 of the Geneva Conventions is silent on the education of children during internal armed conflict, Article 4 (3) (a) of Protocol II states that children shall be provided with the care and aid they need and, in particular, that they shall receive "an education, including religious and moral education, in keeping with the wishes of their parents, or in the absence of parents, of those responsible for their care." Article 24 of the Fourth Geneva Convention, which is applicable to the whole population in a belligerent country, requires states to take the necessary measures "to ensure that children under fifteen, who are orphaned or are separated from their families as a result of the war, are not left to their own resources, and that . . . the exercise of . . . their education [is] facilitated in all circumstances. Their education shall, as far as possible, be entrusted to persons of a similar cultural tradition."[4]

On the importance of equal access to education for girls, analogous application of some standards developed by UNHCR's Executive Committee for refugees is pertinent. The Committee has stressed the high "priority to the education of all refugee children, ensuring equal access for girls, [and] giving due regard to the curriculum of the country of origin." It has also urged UNHCR "to identify educational requirements in the early stages of an emergency so that prompt attention may be given to such needs."[5] Of particular concern are measures to ensure equal opportunity for refugee girls as for boys, equal access of refugee women to adult education, and skills training to ensure that they are able to support themselves and their families.

Beyond education for young people, boys and girls, there is an increasing realization that education should be accessible to adults, especially those needing special training to acquire skills that are required for employment under changing circumstances.

Dilemmas of Emergency Operations

Education in the broader sense requires understanding the context of the emergency operations and the dilemmas inherent in the different perspectives of all those involved, humanitarian aid workers and the recipient communities.

Understanding the problems enhances the prospects for achieving the necessary balance in formulating and implementing programs. The challenges that must be confronted fall into four clusters: the external nature of humanitarian interventions; the relationship between relief activities and endemic problems; the coordination of relief initiatives; and the mixed results of relief operations.[6]

The External Dimension of Emergency Operations

The first challenge concerns the external nature of emergency relief, which represents at the same time its greatest strength and its most serious paradox. Perhaps the most problematic aspect of emergency situations is that international relief operations are undertaken when national leadership and capability have failed and proven themselves unequal to the task of meeting the needs of the population. This is a humiliating fact that generates ambivalent response to external assistance.

In the conflict-generated humanitarian crisis in Southern Sudan, enormous resources from around the world were placed at the disposal of relief agencies in the Sudan, which received assistance in excess of $1 billion in 1984–86. Operation Lifeline Sudan mobilized an estimated $200 million to $300 million in 1989. This aid was widely credited with averting massive starvation, although the number of lives saved was impossible to quantify and many lives were lost despite energetic efforts to save them.

These resources injected an international dimension into what would otherwise have remained domestic issues. However, the value of food aid was probably less important than the benefit of breaking food prices maintained by local merchants at high levels. The presence of international aid personnel in 1989 also helped restrain abuse of civilians and moderated the conduct of the war by both the government and the armed opposition, the Sudan People's Liberation Movement and Army (SPLM/A).

On the negative side, the scale of the operations made the Sudanese feel overwhelmed and excluded from the relief effort. They had no sense of being partners in the effort. The emphasis on external assistance further blurred the need for local people and institutions to view their own efforts as ultimately critical to their survival.

Reflecting on emergency operations in response to the drought-generated famine in Darfur in 1984–85, one observer concluded that "all too often famines are discussed as though the successes and failures of relief were the most important factor in survival." He went on to say that "food aid was not the most important element in surviving the famine. This does not mean that the aid programs were not important, just that they were less important than the people who worked on them and publicize them often believed."[7] The Office of the Emergency Operations in Africa (OEOA) noted in a study of the emergency the remarkable resourcefulness and resilience of the imperiled people themselves, who used survival techniques that were quite successful even though they did not measure up to global standards.

The national humiliation Sudanese authorities associated with OEOA in the 1984–86 emergency operations kept them from more active cooperation with international agencies later in the decade. The backlash may have contributed to the government's unwillingness to acknowledge the scale of the need in 1988. The Islamist regime resisted acknowledging food shortages in 1990–91 partly because it feared that the Christian West was using nongovernmental organizations to infiltrate the Sudan with religious and political agendas. Acknowledging the food shortages, the government feared, would trigger another international relief intervention.

The Disconnect Between Emergency Operations and Long-term Development

The second challenge is the relationship between relief operations and long-term development efforts, beginning with the resolution of the causes of the emergency. During emergencies operational concerns quite naturally take priority. The endemic causes of the crisis— agricultural policies and political, ethnic, and religious tensions—get short shrift. The challenge is to use relief activities as an entry point for addressing matters of long-term consequence: underdevelopment, environmental degradation, human rights violations, and violent conflict. Relief operations need to be sensitive to their context and must try to make maximum use of local human and material resources, social structures, value systems, and survival techniques.

The OEOA initially involved few people with few resources in the Sudan. When food shortages began appearing in 1983–84, Sudanese authorities were slow to respond. Eventually, relief activities, after a sluggish start, mushroomed into a monumental operation that became a source of both appreciation and resentment on the part of the Sudanese. Within six months, the emergency operation had a mammoth structure of its own, involving large numbers of expatriates, more than five hundred local support personnel, offices around the country, and a radio communications network more reliable than the national telephone system. The OEOA mobilized a fleet of almost four hundred trucks and a dozen aircrafts, dwarfing the resources available to the government.

Of course, the crisis was severe enough to justify a warlike mobilization of human and material resources to deliver food to the starving. But that paradoxically became its downside. The relief operation, soon took on a life of its own. One observer described the OEOA as "a top-down, heavy handed operation which was mainly interested in getting more trucks and more radios to equip the trucks with and carrying out a military style operation with very little regard to the actual social, developmental, political . . . sides of the whole equation."[8]

Critics also argued that the international relief effort probably entrenched the causes of the famine itself. Beneficiaries had not been enlisted in the relief activities, and the strategies and technologies employed might have rendered communities more dependent on outside assistance. For example, water pumps required spare parts available only outside the Sudan, and seeds and fertilizers were imported.

So urgent was the need to move supplies into remote areas that Operation Lifeline Sudan became a logistical Mission Impossible. The Khartoum conference in March 1989, which launched Lifeline, set tonnage targets that would haunt the program. With every well-publicized failure of trucks, trains, and barges to reach their destinations, Lifeline seemed a step closer to collapse. It did not reach its end-of-June target of 107,000 tons until October, although by year's end it had moved some 111,654 tons of food and 3,760 tons of nonfood items. While few people knew the nutritional capacities or physical dimensions of a ton, many saw in Lifeline's tonnage reports an indicator—mistaken, as it turned out—of a failed undertaking. By May 1989, a consultant visiting Southern Sudan reported that an overemphasis on tonnages moved had "tended to create a short-term relief mentality that obscures equally important medium-term needs and the opportunity to provide relief in a way that supports, rather than undercuts, longer-term development."[9]

However urgent the logistic challenges of emergency relief operations, the underlying causes of suffering must be addressed if the assistance is to bring about a more food-secure future. In the Sudan, this meant government attention to the causes of chronic food insecurity, including the civil war itself.

Ambivalence Over Coordination

A third challenge is that of coordination. Too little coordination can result in duplication of relief efforts, and too much can hamstring effective operations. Despite widespread agreement on the desirability of coordination, however, few relief agencies welcome the discipline that coordination imposes. Getting the balance right under the pressure of extreme human need is not an easy feat.

An important reason for coordination is the high cost of international relief operations. In the Sudanese case, the levels of assistance were too high to be sustained over an extended period and perhaps even too high for the short term. The OEOA exercised the coordinating role in 1984–86, and Operation Lifeline Sudan did the same later. The United Nations helped alert the international community to the scale of the need, mobilized substantial resources, assisted in the delivery of supplies, served as liaison between aid agencies and Sudanese authorities, and provided basic accountability.

Nevertheless, serious problems developed. Some of the difficulties were essentially technical, such as maintaining accurate, up-to-date, and comparable information from each participant. Other difficulties were political, such as efforts by the authorities, in the name of coordination, to control and constrain relief activities, on occasion using the United Nations for that purpose. Other difficulties reflected interagency rivalries within the UN system. Both technical and political issues were complicated by the civil war.

Mixed Results in Emergency Operations

The fourth challenge concerns the mixed results of emergency operations. When relief activities wind down and the international community moves out, the beneficiaries are supposed to be better equipped for whatever the future may bring. Relief operations often do not measure up to expectations in this respect.

Although international assistance should make political authorities more responsive to the people, it can also reinforce the adverse practices of governments and armed opposition groups, shift the burden of civilian need to the international community, and contribute to dependency on outside assistance.

Outside resources invested in emergency operations in sub-Saharan Africa were significantly higher in the 1980s than in the previous decade. The levels of assistance reflected an increase in conflict-generated need among major recipients of which the Sudan was one. At the same time, however, such aid may have contributed to the continuation of the strife.

As a Kenyan peace activist opined, "The time has come for us in the Horn of Africa, to ask whether the efforts of relief agencies are contributing, indirectly or even remotely, to an escalation of the wars or to a peaceful resolution of the conflicts."[10] Lifeline's preoccupation with logistics helps account for its inattention to the conflict. On the positive side, however, Lifeline brought the war to the attention of the world and triggered efforts to find a solution, which, though protracted, were sustained and eventually bore fruits two decades later.

The interplay of political realities and humanitarian objectives contributes to the mixed results of relief activities. In the drought-related crisis, the Sudanese government's reluctance to acknowledge the famine was clearly political. Donor governments, in particular the United States, were perceived to be boosting the Nimeiri regime, which became an ally of the West on a number of strategic issues. Most Western donors were reluctant to press Khartoum to deal with human need when the regime was not prepared to do so. But eventually the West, led by the United States, pressured Nimeiri into declaring a state of emergency and inviting international assistance.

Lifeline was also highly political. "Relief is not a value-free operation," observed the Sudanese relief and resettlement commissioner. "It is based on the interest of the countries involved. It has a cultural and a religious dimension. It is a network of sometimes conflicting interests and forces pushing toward different goals, though dressed up in the same garb."[11]

Less widely known was the questioning by the Sudanese themselves of the usefulness of relief efforts. One professor lamented, "We should have been left alone; we were dying anyway." Disturbed by what he saw as an arrogant and obtrusive foreign relief community in Khartoum, he also felt that relief operations reinforced Nimeiri's regime, gave him international legitimacy, and perhaps prolonged his last days with gross violations of the very human rights he himself advocated.[12]

Both outside and inside the Sudan, then, informed persons of compassion and conscience found themselves profoundly ambivalent about emergency operations. Aid interventions, they found, demonstrate that humanitarian initiatives, however compelling on moral grounds, exercise no compelling authority when political or military forces are otherwise inclined. In addressing one problem successfully, relief activities may also create or compound others.

At the same time, the Sudan experience suggests that humanitarian initiatives, creatively managed, can be an influential force for positive change in their own right. Interventions can be made more proportionate and precise and their side effects minimized. Costs can be contained, although relief efforts will never be cheap. If the complex motives and mixed results of humanitarian aid are acknowledged, the international community can accentuate its humanitarian purposes and achieve the best possible results for those in need, not only for the immediate relief purposes, but also for addressing the deeper causes of the crises.

Cultural Contextualization of Education

Relief emergency operations cannot succeed in the long run without due regard to the cultural context. Culture is an embodiment of values that have evolved over a long period of time and have become crystallized into normative frameworks and institutionalized patterns of behavior. Cultures and the values they engender are what give human beings their innate dignity and social cohesion. While cultures are dynamic and change with the imperatives of experience, they also provide predictable standards for prescriptive behavior. The result is a social order that reinforces perspectives, expectations, and conformity. Change then takes the nature of discrete reforms, with occasional revolutionary zeal, prompted primarily by inspired charismatic leaders or by crises of grave magnitude. Otherwise, life is evolutionary and relatively predictable.

Applying these principles to the African scene, it is obvious that the continent has suffered a trauma that has severely damaged its indigenous structures, perspectives, and behavior patterns. Politically, economically, culturally, and spiritually, the peoples of Africa were assumed to be in a void that had to be filled by imported concepts, values, and aspirations for self-enhancement through the emulation of the supposedly superior outsiders and their cultures. Rather than see development as an evolution or a process of self-improvement from within, Africans began to see it as a commodity that had to be imported, with indispensable assistance from outsiders.In his message to a symposium organized by Dr. Kevin Cahill, "Traditions, Values, and Humanitarian Action," the then Secretary-General of the United Nations, Kofi Annan, had this to say on the subject:

Traditions are what distinguish each human society from all the others. They are what each society brings to the great banquet of human diversity. Indeed, they are the essence of that diversity itself, which is what makes the human species such a rich and splendid one to belong to.

The tragedy of human history is that so often people have allowed diversity to drive them apart instead of bringing them together, have interpreted their traditions exclusively, and taken refuge within them. Too often, in the name of tradition, anathemas have been pronounced and wars have been fought.

Values are what enable us to overcome those divisions—to approach one another with confidence and curiosity rather than fear and suspicion, to learn from each other, to respect each other's traditions, to cherish our diversity.

In short, if traditions can and should be kept distinct, values need to be shared. And first among our shared human values must be the humanitarian instinct, the instinct that drives us to help our fellow human beings in their hour of need, no matter how different from us they may be.[13]

Such help, however, must give deference to cultural mores of those in need. Dr. Cahill himself testified to the importance of respecting indigenous cultures and values and the impact that had on his humanitarian medical services in the Sudan:

In the early 1960s, I worked for many months as a physician in the Southern Sudan. It was a time of great social unrest and revolution in an area long isolated from the impact of modernity. The missionaries, who provided the only health and educational services available, were ejected shortly after my arrival. I found myself the only doctor within hundreds of miles of roadless, swampy land, the Nilotic Sudd, home of the Dinka, Nuer, and Shilluk tribes.

Offering basic emergency medical services exposed me to customs and practices of which, to that time, I was utterly ignorant. They were not based on our Western traditions and values and, initially, seemed to me the relics of a primitive culture. Over months, however, I came to respect the strength and beauty of their ways and beliefs.[14]

As emergency situations often result from identity-related conflicts, an aspect of education in the broader sense of the concept must be to understand the dynamics involved in the conflict of identities and how to address and resolve them. This is essentially a challenge of nation-building, which raises the question of how to foster a sense of common identity and purpose while also recognizing and building on the distinctive features of the various ethnic and cultural elements that constitute the nation-state. This applies to learning as an integral component of the social and cultural context of a particular people, a means of acquiring knowledge about their overriding moral and cultural values and the various ways by which they strive to approximate their personal and societal goals, spiritual or material. Both the means (learning) and the end (knowledge) are reflections not only of a people and their ancestral heritage, but also of how the people relate to their environment and utilize the resources available in their natural setting. They constitute both a source of identity and a means of survival. I illustrate with the culture with which I am most familiar, that of the Dinka of the Sudan.

Dinka Concept of Knowledge

As noted, knowledge comprises the material components of the culture of survival and the moral, social, and spiritual values that nourish the essence of their humanity in context. The Dinka have a myth that explains their functional approach to knowledge, embodying the material foundation of their culture. According to the myth, the Dinka chose the cow in preference to the thing called "what," which God had offered them as an alternative to the cow, but which the Dinka dismissed without even exploring what it meant.[15] As one elder explained it,

> God asked man, "Which one shall I give you, black man; there is the cow and the thing called 'what,' which of the two would you like?" The man said, "I do not want 'what.'" Then God said, "But 'what' is better than the cow!" The man said "no!" God said, "If you like the cow, you better taste its milk first before you choose it finally." The man squeezed some

milk into his hand, tasted it, and said, "Let us have the milk and never see 'what.'"[16]

"What" was later given to other peoples and presumably became the source of their inquisitiveness, scientific invention, and modern education, while the Dinka remained immersed in a cattle-dominated culture, tradition-bound education, and backward-looking indigenous knowledge, whose value in the modern context they are beginning to question.

The Dinka term for knowing is *ngic*. I doubt that it can be appropriately translated as knowledge because, in itself, *ngic* does not imply a notion of substantive content, but rather a condition of knowing something. Indeed, as a noun, the word *ngic* is always used in conjunction with an object of knowledge, that is, knowledge of something as opposed to knowledge as an abstract generic concept.

The most common reference to knowledge is *nginy e wel*, literally "knowing the words." A person who is knowledgeable in that sense is *raan ngic wel*. But what is meant by this expression is more than an accumulation of factual information or ability to articulate. It also entails making an effective use of the information to attain desirable moral goals. To say that "a man knows the words" means that he is well-informed, articulate, wise, considerate, prudent, tactful, diplomatic, and persuasive.

Knowing the words should be distinguished from "knowing to speak," *nginy e jam*. Although verbal talent is appreciated as a skill, unless it is qualified by a positive moral content, it may carry a negative connotation; a person with such a talent could cleverly sway people without moral cause or content. Conversely, a person who does not know how to speak, *raan kuc jam*, may be morally lacking or may be merely disadvantaged, and yet be morally superior to a person who is well-spoken without the wisdom normally implied in "knowing the words." In a case I witnessed, a man who was notorious as an informant of the security forces in a war situation, as a result of which many people were arrested, tortured, and killed, bragged about his rhetorical abilities. An elder in the meeting responded by saying, "No one doubts your ability with words; it is the state of your heart that people doubt."

A person who does not know the words, *raan kuc wel*, is not only one who cannot express himself or herself well, but one who is also lacking in the propriety of mannerism, conduct, or prudence. Such a person may be stereotyped as unwise, reckless, aggressive, obstreperous, and insensitive.

It is apparent from these illustrations that there is a moral and ethical content to "knowing the words." The range of areas covered by knowledge of the words is extensive and not easy to define. It covers such religious affairs as the myths of creation or the legends of the ancestors, family and tribal history, social norms and mores, and the mystic traditions that explain the spiritual relationship of humanity to other elements of the universe.

There are other areas of knowledge generally referred to as *nginy e kang*, "knowledge of things," or *nginy e luoi*, "knowledge of work," which may not enjoy the same degree of moral and spiritual reverence as "knowledge of the words," but that are nonetheless vital to life and physical survival. These cover a wide range of practical knowledge and artistic skills that are essential to livelihood, well-being, and cultural nourishment. Medical expertise ranges from spiritual cure to the use of herbs, the setting of bones, and the performance of a variety of operations, some of them quite delicate. Expertise in agriculture and animal husbandry entails knowledge of seasonal changes as indicated by the behavior of the moon and the stars, climatic and weather conditions, sometimes predicted with surprising accuracy, the suitability of certain crops to certain types of soil, methods of seeding and weeding, and seasonal migration with herds in search of appropriate grass and water. Other skills include building, weaving, leather making, pottery making, hunting, warring, singing, and dancing. Indeed, areas of skilled knowledge are as inexhaustible as there are situations requiring practical know-how to sustain a culturally oriented pattern of life. It is indeed in these areas of knowledge that survival mechanisms and techniques in emergency situations can be found.

While studies of traditional society have focused mainly on those areas covered by "knowledge of the words," what we might call the humanities, the wide array of Dinka knowledge of skills needed for physical survival has not drawn much scholarly attention. This is not merely an oversight by students of Dinka society, but results from the emphasis the Dinka themselves place on the concept of knowledge relating to the words—philosophic or other values—that tends to overshadow information on the more practical side of their culture.

Dinka Moral, Social, and Spiritual Values

If education among the Dinka is essentially normative, centered on learning their moral code of conduct, it is of pivotal importance to know the values that comprise this code. Three sets of principles provide the pillars of the code: continuity through the ancestral line, a form of immortality; idealized human relations centered on unity, harmony and cooperation; and respect based on individual and collective dignity. Anyone working with the Dinka, whether in the area of humanitarian assistance or development, would be well advised to be mindful of these values as vital elements of the cultural and social environment in which they work.

To the Dinka, the family is the backbone of society and the foundation of its social, moral, and spiritual or religious values. The overriding goal of every Dinka is to marry and produce children, especially sons, "to keep the head upright," *koc nom* (or *nhom*), after death. Although the Dinka believe in some form of continued existence that conceptually projects this worldly life into the hereafter, death for them is an end from which the only salvation is continuity through posterity.

As Godfrey Lienhardt, the British anthropologist who studied the Dinka, notes, "Dinka fear to die without male issue, in whom the survival of their names—the only kind of immortality they know—will be assured."[17] When a man dies without issue to carry on his name, members of his family are under a moral obligation to marry a woman for him, to live with a relative and beget children to his name. Equally, a man who dies leaving behind a widow of childbearing age devolves a moral obligation on his kinsmen to have one of them cohabit with her to continue bearing children to his name.

This worldly orientation of religion has the effect of making the Dinka intensely religious, with high standards of moral values essential to physical and psychological well-being. The consequences of good and evil for the Dinka are not deferred to the afterlife; they are here and now. And every illness or misfortune is believed to have some moral cause in the actions of the victim, a close relative, or an intrusive evildoer. According to the anthropologists Brenda Seligman and Charles Seligman, "The Dinka, and their kindred the Nuer, are by far the most religious people in the Sudan."[18] As Lienhardt observes, "Divinity [God] is held ultimately to reveal truth and falsehood, and in doing so provides a sanction for justice between men. Cruelty, lying, cheating, and all other forms of injustice are hated by Divinity, and the Dinka suppose that, in some way, if concealed by men, they will be revealed by him."[19] When a man has a grievance for which he needs divine intervention, all he needs to do is invoke the existence of God, "Nhialic ato thin"—God exists. And, as Lienhardt notes, "It is a serious matter when a man calls on Divinity to judge between him and another, so serious that only a fool would take the risk involved if he knew he was in the wrong, and to call upon Divinity as witness gives the man who does so an initial presumption of being in the right."[20] Chief Thon Wai reflected this unwavering faith in divine justice when he said:

> Even if a right is hidden, God will always uncover the right of a person. It doesn't matter how much it might be covered; even if the covering be heaped as high as this house and the right is there, it will appear. It may be covered for ten years, and God will uncover it for ten years, until it reappears. . . . If a man is not given his right, God never loses sight of the right.[21]

Despite the warlike profile of the Dinka, their moral values emphasize the ideals of peace, unity, harmony, persuasiveness, and cooperation. These values are highly institutionalized and expressed in a concept known as *cieng*, which is fundamental to Dinka moral and civic order. Lienhardt wrote of *cieng*: "The Dinka . . . have notions . . . of what their society ought, ideally, to be like. They have a word, *cieng* baai [baai meaning home, village, community, or country], which used as a verb has the sense of 'to look after' or 'to order' and in its noun form means 'the custom' or 'the rule.'"[22] Father Nebel, who lived among the Dinka and became a leading expert on their culture and way of life, translated morals as "good cieng" and benefactor as a man who knows and acts in accordance with cieng. He also translated cieng to mean "behavior," "habit," "nature of," or "custom."[23]

At the core of cieng are such human values as dignity and integrity, honor and respect, loyalty and piety, and the power of persuasiveness. Cieng does not merely advocate unity and harmony through attuning individual interests to the interests of others; it requires assisting one's fellowmen. Good cieng is opposed to coercion and violence, for solidarity, harmony, and mutual cooperation are fittingly achieved voluntarily and by persuasion.

Cieng has the sanctity of a moral order not only inherited from the ancestors, who had in turn received it from God, but also fortified and policed by them. Failure to adhere to its principles is not only disapproved of as an antisocial act warranting punishment, but, more importantly, as a violation of the moral code that may invite a spiritual curse—illness or death, according to the gravity of the violation. Conversely, a distinguished adherence to the ideals of cieng receives temporal and spiritual rewards. Although cieng is a concept with roots in the heritage of the ancestors who still sanction adherence to its principles, it is largely an aspiration that is only partially adhered to and, indeed, often negated. Hence, it can be improved upon even through innovation.

The contradiction between the requirements of cieng and the violent reputation of the Dinka can be explained in terms of the gap between the ideal and the real, institutionally manifest in the differences between generational roles. While elders strive to live by the ideals, the young warriors find self-fulfillment, social recognition, and dignity in their valor, fighting ability, and defensive solidarity. Consequently, they tend to overindulge in militancy, often responding violently to the slightest perception of insult or disrespect, provoking wars that must then be fought by all. Nevertheless, frequent and pervasive as it is, warfare reflects a negation of the ideals, an alternative that should only be resorted to when peaceful methods have failed.

By the same token, chiefs, even when young, must be men of peace. One chief, reacting to the assertion that, traditionally, force was the deterrent behind Dinka social order, articulated the delicate balance between the violence of youth and the peacemaking role of chiefs: "It is true, there was force. People killed one another and those who could defeat people in battle were avoided with respect. But people lived by the way God had given them. There were the Chiefs of the Sacred Spear. If anything went wrong, they would come to stop the . . . fighting and settle the matter without blood. Men [chiefs] of the [sacred] spear were against bloodshed."[24] And in the words of another chief, "There was the power of words. It was a way of life with its great leaders . . . not a way of life of the power of the arm."[25]

While it was not always easy for the elders to control the overzealous warriors, it was a clearly established principle that the warriors should adhere to the will of their chiefs and elders. When they had to confront aggression and justifiably go to war, they were blessed by their spiritual leaders, chiefs, instructed on the ethics of warfare, and, trusting in the justice of their cause, counted on their ancestral spirits and deities to ensure their victory against the forces of evil. Among the ethical principles of warfare were that the enemy must not be ambushed or killed outside the battlefield and that a fallen warrior covered by a woman for protection [women accompany men in battle primarily to help the wounded] must be spared, as harming women and children in war was strictly forbidden.

It is particularly noteworthy that despite the lack of police or military forces, civil order was maintained with a very low level of crime, other than those incidents associated with honorable fighting or pursuit of self-help forms of justice. Major Court Treatt, who traveled in Dinkaland in the late 1920s, described the Dinka as "a gentleman" who "possesses a high sense of honor, rarely telling a lie," and with "a rare dignity of bearing and outlook."[26] And Major Titherington, who served as a colonial administrator among the Dinka also in the late 1920s, wrote of "the higher moral sense which is so striking in the [Dinka]. Deliberate murder—as distinct from killing in a fair fight—is extremely rare; pure theft—as opposed to the lifting of cattle by force or stealth after a dispute about rightful ownership—is unknown; a man's word is his bond, and on rare occasions when a man is asked to swear, his oath is accepted as a matter of course."[27] Sir Gawain Bell, who served as District Commissioner among the Ngok Dinka, observed, "I can't remember that we ever had a serious crime in that part of the District. Among the Bagara [Arabs] . . . there was a good deal of serious crime: murders and so forth; and the same applied to the Hamar in the North. . . . The Ngok were a particularly law-abiding people."[28]

One of the ways in which Dinka culture sustains a level of conformity and continuity is by providing alternatives to dignity that accommodate and institutionalize even elements of what would otherwise be a violation of the norms. An initiated man is *adheng*, a "gentleman"; his virtue is *dheeng*. But initiation is only a key or a point of entry to the complex values of *dheeng* and their varied avenues to individual and social dignity. *Dheeng* is a word of multiple meanings, all of them positive. As a noun, it means nobility, beauty, handsomeness, elegance, charm, grace, gentleness, hospitality, generosity, good manners, discretion, and kindness. Except in prayer or on certain religious occasions, singing and dancing are *dheeng*. Personal decoration, initiation ceremonies, celebration of marriages, the display of "personality oxen"—indeed, any demonstrations of aesthetic value are considered *dheeng*. The social background of a man, his physical appearance and bearing, the way he carries himself, walks, talks, eats, or dresses, and the way he behaves toward his fellowmen are all factors in determining his *dheeng*.

From its various meanings, one can discern at least three kinds of *dheeng* or dignity. The first derives from birth or marriage into a family with already established status. The second is the status people acquire through material resources and social responsiveness, which is measured not only in terms of generosity and hospitality, but also by personal integrity and responsible conduct. The third is more sensual in nature and stems from physical attractiveness and various forms of artistic display.

To the Dinka, power and wealth must serve moral and social ends, or else they do not confer *dheeng* on the holder. The word for chief, *beny*, also means "rich" or "wealthy," but a man of wealth who is stingy or too frugal is *ayur*, the opposite of *adheng*, and his indignity is *yuur*. Conversely, a man of modest means who is generous and hospitable is praised as *adheng*, and even as *beny*. On the other hand, a man who is exaggeratedly generous or hospitable far beyond his means is considered vain and a showoff, *alueth* (a word that also means "liar"); although he is not despised in the way a stingy person with means would be, his performance falls short of *dheeng*. The ideal behavior is for a person to have the means and to display a social consciousness commensurate with his means. That is *adheng* in the ideal sense.

A Dinka chief's legitimacy rests in the normative framework of a spiritual leader whose power rests on divine enlightenment and wisdom. In his installation ceremonies, people raise his right hand toward the sky, symbolizing the will of his people and acceptance by the divine powers above. In the case of aggression against his tribe, when force is necessary to stop force, the chief should pray for victory far away from the battlefield. In a war between his own factions, he should not take sides, but rather should pray for peace, draw a symbolic line, and place his sacred spear upon it, while willing that the group that crosses it to attack will suffer heavy casualties. In order to reconcile his people, the chief should be a model of virtue, righteousness, and, in Dinka terms, "a man with a cool heart," who must depend on persuasion and consensus rather than on coercion and dictation. The word for "court" or "trial," *luk*, also means "to persuade." Godfrey Lienhardt wrote on this aspect of the Dinka:

> I suppose anyone would agree that one of the most decisive marks of a society we should call in a spiritual sense "civilized" is a highly developed sense and practice of justice, and here, the Nilotics, with their intense respect for the personal independence and dignity of themselves and of others, may be superior to societies more civilized in the material sense. . . . The Dinka and Nuer are a warlike people, and have never been slow to assert their rights as they see them by physical force. Yet, if one sees

Dinka trying to resolve a dispute, according to their own customary law, there is often a reasonableness and a gentleness in their demeanor, a courtesy and a quietness in the speech of those elder men superior in status and wisdom, an attempt to get at the whole truth of the situation before them.[29]

Relief operations should not and cannot function in a cultural vacuum or oblivious to the context of their humanitarian work. To be informed about this context is to be educated in a functional sense that ensures the cultural and social contextualization of education in the broad meaning of the concept.

Understanding the Root Causes of Conflict

Emergency situations mostly result from identity-related genocidal conflicts and mass atrocities usually directed against groups identified by nationality, race, ethnicity, or religion, the elements specified in the 1948 Convention on the Prevention and Punishment of the Crime of Genocide. It must, however, be emphasized that it is not the mere differences, real or perceived, that generate conflict, but the implications of those differences in terms of access to power, wealth, services, employment, development opportunities, and the enjoyment of citizenship rights. Some qualify as members of an in-group that enjoys all the rights of citizenship, while others are discriminated, marginalized, excluded, and denied the dignity of citizenship and the rights accruing from it. Their status as citizens becomes largely of paper value.

In the past, when these disadvantaged groups were isolated and lacking the confidence and the capacity to resist, they silently succumbed to their plight as ordained by the dominant powers. Today, with the global consciousness of universal human rights, including the rights of minorities and indigenous peoples the world over, such a demeaning status cannot be sustained without resistance. However, when the aggrieved resist, sometimes using violent means as a last resort, they provoke the more powerful forces of the status quo to respond with devastating consequences that can escalate to genocidal levels.

This assessment can become a powerful tool for prescribing remedies and curing chronic societal ills that can generate genocidal crises. The critical step then becomes one of identifying the factors in a system that account for acute cleavages and disparities in the management of diversity and to seek ways of reducing and even eliminating the gross inequities that can generate resistance, insurgency, and genocidal counterinsurgency.

Indeed, one of the major challenges to conflict resolution and nation building is the management of diversity based on these elements of identity. In the case of Africa, given the fact that most modern States were carved out in a manner that separated some groups and brought together other groups, diversity is a pervasive reality of the African state. Nor is this diversity only a function of the more easily identifiable racial, ethnic, "tribal," or religious differences. As Somalia has so dramatically demonstrated, even differences among clans, and, indeed, lineages or families in a society that is otherwise homogeneous can be most violent and threatening to the survival of the state and society. If ethnic conflict, whether broadly or narrowly defined, is an increasing phenomenon in the world today, especially after the cold war, and if governance is conflict management, then managing or resolving identity conflicts must be high on the agenda of responsible sovereignty. This is particularly the case considering that identity cleavages, if not bridged creatively and equitably, can generate violent conflicts that can escalate to genocidal levels.

There are two sets of discrepancies that are central to the management of identity cleavages. One has to do with the degree to which the subjective factors of self-identification match the objective elements of the claimed identity. Under normal circumstances, this is a personal matter that should not concern others. The other set of discrepancies has to do with the degree to which exclusive individual or group identities are reflected or represented in the definition of the collective national identity framework. The discrepancy between the exclusive identities and the collective national identity makes the issue of identity, in both its subjectivities and its objectivities, a matter of public policy and, therefore, scrutiny. To the extent that the issue impinges on the interests of other citizens, identity enters the public domain and ceases to be purely personal or exclusive to a group. If an exclusive identity conflicts with the requirements of national unity in a framework of diverse identity, then a need arises either to remove the divisive elements and redefine the national identity framework to be all-inclusive, or to design a system of coexistence among the diverse groups through constitutional arrangements that accommodate at least the more significant diversities, or, as a third option, to allow the diverse parts to go their separate ways.

The Quest for Constructive Management of Diversity

In a brainstorming meeting at the United States Institute of Peace (USIP) in 2003 to discuss the project that resulted in my book *Identity, Diversity, and Constitutionalism in Africa*, a program officer at USIP quoted from *Our America* by José Martí, a Cuban independence hero, who wrote, "To govern well, one must see things as they are. And the able governor in America is not the one who knows how to govern the Germans or the French; one must know the elements that compose one's own country, and how to bring them together, using methods and institutions originating within the country, to reach that desirable state where each person can attain self-realization and all may enjoy the abundance that nature has bestowed on everyone in the nation."[30] The challenge confronting African countries as they strive to govern their people within the complex racial, ethnic, religious, and cultural realities of each country could not have been better articulated.

It is becoming recognized that all African countries strive in varying ways and degrees to transcend the simplistic Eurocentric model constitutions and principles of constitutionalism that assumed a degree of homogeneity with hardly any regard to the specificities of the African context, its cultural values, institutions and patterns of behavior. Two main questions need to be addressed in searching for contextualized solutions to Africa's constitutional arrangements: to what extent do African constitutions recognize, respect, and indeed, utilize the regional, ethnic, cultural, and religious diversities of the country concerned? And to what extent do they connect to indigenous mechanisms or build on the cultural values and institutions of the country?

These questions also trigger other sets of related questions: Is there one universal or global model of constitutions and constitutionalism? Is Africa bound to follow the Eurocentric models which the colonial powers bequeathed to them at independence? In other words, are Africans predestined to follow a linear approach in which they are at an early stage of evolution of modernization? These questions may sound rhetorical, but they do raise serious issues that are debatable. Even if there were a broad consensus on the normative response to the questions, the realization or application of the principles involved would raise practical questions not so much on what is desirable but rather on what is doable.

The subcategories of the overarching questions that the conceptual framework also presents are how managing diversity and contextualizing constitutionalism are translated in several areas of governance: conflict prevention, management, and resolution; participatory democracy; respect for fundamental rights and civil liberties; self-reliant development; environmental protection; and gender equality. These issues have both global and local dimensions and parameters.

An overriding normative principle is a concept of self-determination that entitles individuals and groups to play a pivotal role in the management of their own affairs. It is ultimately a principle of human dignity for groups and individuals at all levels of political, economic, cultural, and social organization or structures, from the national down to the local, with the individual as the final unit. The dilemma for constitutionalism in Africa is inherent in the fact that self-determination, perceived as a group right that may be exercisable through secession as an option threatens the unity and integrity of the nation. It can be assumed, however, that a state that discharges its responsibilities toward its citizens, individuals and groups, through constructive management of diversity, democratic participation, equitable sharing of wealth and development opportunity, and respect for fundamental human rights and civil liberties, is not likely to be threatened by the fragmenting demand for self-determination. On the other hand, a state that fails grossly to meet these basic objectives of good governance can expect the aggrieved groups to demand self-determination in its separatist version leading to secession.

Although it is debatable whether African constitutions failed mostly because they were founded on alien models or because African governments did not adhere to the normative principles of the European models, the fact is that in either case, they did not build on the strengths of their own value-systems and institutional structures. Even if African leaders wanted to follow the ideals of Western constitutionalism, without being grounded in their indigenous cultures, values, structures, and processes, they would be building castles on sand. There can be no question that Africa stands to benefit from cross-cultural fertilization, but that should mean what the word implies: synergizing the positive aspects of the interactive culture and related value-systems. An article of faith behind this approach is that there is an essence in African traditions that can make a forward-looking constitution. The challenge, then, is to find out what these values are and how they can be interpreted to synergize with received Western ideas and practices to facilitate a constructive and productive transitional integration between tradition and modernity.

In the indigenous African system, the autonomous structure of governance guaranteed a degree of accommodation for diversity. However, the centralization of government in the modern state context fundamentally undermined the indigenous power structures and processes. An area in which the indigenous system deserves criticism is gender equality and the participation of women. But rather than simply criticize traditional inequalities, it is more constructive to understand the basis of the indigenous system, the changes in goals and strategies the modern context presents, and the need to adjust to those changes in pursuit of human dignity on an equal basis for males and females.

In this regard, it is important to recognize that there are universal minimum standards that are related to the universality of human dignity, defined as the broadest shaping and sharing of all values: material, social, moral and spiritual. Those universal norms are not specific to any one country or culture but are indeed the common legacy of humanity. In other words, the existence of these universal norms denies the exclusivity of the ideals of constitutionalism to any culture or civilization. What each country or culture is called upon to do is to identify these universals and to give them local grounding and legitimacy. The challenge then becomes a dialectic approach of how to extrapolate those common, universal elements in the specificity of the local context. With those universal principles as the benchmarks, the performance of each country can then be evaluated.

Applying these benchmarks to the specific areas of concern then becomes one of using local cultural values and institutional practices in pursuit of the ideals that are cross-culturally synthesized and enriched. For instance, given the fact that conflicts are a fact of life in all contexts, the issue is not to fantasize or romanticize a situation divested of conflict, but rather how to manage and mediate conflicts to reduce or avoid their most destructive aspects, often manifested in violence. Human rights are inherent in every society's pursuit of human dignity, but the precise form such pursuit takes leaves a gap between the universal ideals and the practice of individual societies. The same can be said about democratic participation. Likewise, development as a principle of self-enhancement from within is a process experienced by all societies, but with differing degrees of success. Sensitivity to the environment and attitudes toward gender roles are also culturally specific, and while all cultures have their own ways of managing the challenges involved, they stand to gain from universal pursuits with local grounding.

All said and done, no African country appears to be performing with unqualified success in developing a system of constitutionalism that manages diversity constructively and builds on indigenous values and institutions. Nor is there an expectation that such a system can be put in place that can be judged as an unqualified success. What is aimed at, and can indeed be done, is to make genuine efforts to identify elements of the indigenous cultural values and institutions that can be built upon, and integrated with identifiable universal norms of constitutionalism and good governance. Perhaps the most that can be accomplished is a process of trial and error and of learning from practical experience. And, needless to say, it is also a function of education that is pertinent to understanding the challenges of emergency situations.

Ten Principles on Negotiations

A focus on negotiating skills should be part of the educational process developed for those suffering from trauma and loss that are inevitable in post conflict and disaster situations. At every level—for children whose lives have been utterly disrupted, for teenagers trying to reenter society after serving as child soldiers, for girls and young women seeking to establish an identity with dignity in settings where normal family support structures have evaporated—education should emphasize negotiation as an alternative to violence.

Beyond the conventional notions of teaching children and young people, education in the broader sense I have adopted in this chapter would require all those involved in the emergency operations and who must interact with a wide range of actors—representatives of governmental and nongovernmental organizations, relief workers, aid recipients and members of civil society—to develop as well the art or skill of negotiating complex and highly ambivalent relations. As this chapter has tried to explain, these relations are often tense and conflictual, even as they must build on cooperation and mutual support. Those complexities are often further complicated by cultural differences and cross-cultural perspectives on the issues involved and how to manage diversity constructively. This calls for appreciating for both local and universal principles of negotiations.

I believe the principles outlined in this chapter are pertinent to the framework of education in the broad sense that I have adopted and in the context of conflict management associated with emergency situations. These guidelines derive from personal experiences and are rooted in values, norms and mores that emanate from a specific African family and cultural background among the Dinka of the Sudan.

They cover experiences in interpersonal relations, third-party mediation, and diplomatic negotiations, with overlaps. Although personal and rooted in the Dinka, Sudanese, and African cultural contexts, these principles represent values that can claim universal validity, despite cross-cultural variations on the details and their applicability. They also constitute a challenge in negotiating human relations in emergency situations, whether in bilateral talks between individuals, in third-party mediation, or in diplomatic dialogues.

Interpersonal Relations

- Principle One: Rights and wrongs, though seldom equal, are rarely one-sided. Even when you feel sure that you are in the right, you must not only strive to fit yourself into the shoes of the other side, but must also make the other side recognize that you are genuinely interested in his or her point of view.
- Principle Two: the Dinka believe that it is unhealthy to keep grievances "in the stomach" or "in the heart." "Talking it out"[31] is considered to be not only the best way to resolve differences or grievances, but is also essential for one's mental and even physical health. Often, "What is not said is what divides," to use the words of an article on that theme.[32]
- Principle Three: In the Dinka and indeed African value-system, face-saving is crucial to resolving conflicts. One must avoid saying anything that is humiliating to the other side, and, where possible, it is advisable to show deference, even to an adversary, provided it is not cheap flattery.
- Principle Four: It is important to listen very carefully and allow the other party to say all that she/he considers significant or relevant. Resolving differences is not a game of "wits" or "cleverness," but of addressing the genuine concerns of the parties in conflict. In the Dinka fox stories, the cleverness of the fox eventually turns against the fox. Ideally, resolutions must have an element of give and take, although the distribution should be proportional to the equations of the rights and wrongs involved. In assessing the outcome of a negotiated settlement of a dispute, it is unwise to boast

of victory, for that implies defeat for the other side and therefore an unsatisfactory result.

- Principle Five: Historical memory of the relations gives depth to the perspectives of the parties and the issues involved, but one must avoid aggravating the situation with negative recollections and emphases and should instead reinforce constructive dialogue with positive recollections or interpretations of past events, without, of course, distorting the facts.

Third Party Mediation

- Principle Six: The mediator must be seen to be impartial, but where there is reason to believe that he or she is closer to one side in whatever capacity, the mediator must reach out to the more distant party, although this should not be at the cost of fairness to the party closer to the mediator. Impartiality does not mean having no position on the issues in dispute, even though voicing opinions should be carefully coached to maximize the bridging role and promote mutual understanding.
- Principle Seven: The mediator must listen very patiently to both parties, and even when there are obvious flaws in what is said, the mediator must appear to give due weight to each party's point of view. The popular view that in the indigenous African system of dispute settlement, people sat under the tree and talked until they reached a consensus, reflects a broadly shared African normative behavior. Where explaining the opponent's view on a specific issue might facilitate the bridging process, the mediator should intercede to offer an explanation as part of consensus building.
- Principle Eight: While the wisdom of words and the ability to persuade are important, leverage is pivotal. This means that the mediator must have, or be believed to have, the ability to support the process with incentives or threats of negative consequences, according to the equations of the responsibility for the success or failure of the negotiations. In the past, the spiritual powers of divine leadership provided the required leverage. In the modern context, influencing the balance of power to create a "mutually hurting

stalemate" and help advance the process of "ripening for resolution," to borrow the famous words of William Zartman, is part of the leverage that can effectively facilitate the mediator's task.[33]

Diplomatic Negotiations

- Principle Nine: Diplomatic negotiations combine elements of both interpersonal relations and third party mediation in that the negotiator represents his/her government and in a sense combines negotiating with mediating between the respective governments involved. Discretion and creativity in adapting the official position to the dynamics of the situation with a degree of flexibility is critical to the prospects of successful bridging.
- Principle Ten: While the tendency of the negotiators is to see the outcome of their efforts in terms of winning or losing, especially for domestic consumption, the desired outcome should be one in which neither side sees itself as a total winner or loser, except where the rights and wrongs involved are incontrovertibly clear. The win-win formula should be the objective and whatever the equations of winning or losing in the mediated or negotiated outcome, as noted in Principle Four, neither side should boast about winning and by implication humiliate the other side as a loser. There must be a degree of parity in both sides winning or losing.

Conclusion

By way of summing up, I have tried in this chapter to address the subject of education in emergencies very broadly, not limited to the formal sense the word is normally used. I have also tried to address the challenges emergency situations present with respect not only to access to basic education and institutions of learning generally, but also to other parameters of the emergency context. These include understanding the dilemmas emergencies present for those called upon from the outside to assist, as well as for the recipient countries. Included also is the need for appreciating and respecting the cultural context and the values of the people whom emergencies usually reduce to a level of degradation that makes respecting them a difficult moral challenge. Furthermore, while the nature of emergencies emphasizes responding to immediate needs the symptoms—it is important to understand the root causes, if sustainable remedies are to be found. Because the conflicts that generate emergencies are often identity-related, there is a commensurate need to understand the dynamics of identification, its subjective and objective dimensions, and the tendency to exclude, denigrate, discriminate, and deny others the dignity of common humanity. The remedy is inherent in the analysis, and that is to manage diversity constructively and to turn it from a source of conflict to an enriching resource. This is essentially the challenge for constitutionalism and governance in divided nations. I conclude the chapter with ten principles concerning negotiations between individuals, groups, or intergovernmental representatives, all of which interplay in emergency operations involving local, national, and international adversaries and partners.

I conclude this book with a poem. In rhyming stanzas, Maya Angelou captures the unique essence of children. She reminds us why we must struggle to preserve a future for those beautiful, innocent children who are in mortal danger of losing the potential to grow into caring adults. Education is the ultimate liberator of precious imaginative minds. Accompanying the poem was a letter, and I include a few sentences as her own introduction:

> I have written this poem, praising the indomitable spirit of children, their oftentimes unspoken decision to survive, and even thrive, under conditions which appear insurmountable. I offer it with my most profound desire to serve our children, all our children, everywhere.

Life Doesn't Frighten Me

MAYA ANGELOU

Shadows on the wall
Noises down the hall
Life doesn't frighten me at all.

Bad dogs barking loud
Big ghosts in a cloud
Life doesn't frighten me at all.

Mean old Mother Goose
Lions on the loose
They don't frighten me at all.

Dragons breathing flame
On my counterpane
That doesn't frighten me at all.

I go boo
Make them shoo
I make fun
Way they run
I won't cry
So they fly
I just smile
I just smile
They go wild
Life doesn't frighten me at all.

Tough guys fight
All alone at night
Life doesn't frighten me at all.

Panthers in the park
Strangers in the dark
No, they don't frighten me at all.

That new classroom where
Boys all pull my hair
(Kissy little girls
With their hair in curls)
They don't frighten me at all.

Don't show me frogs and snakes
And listen for my scream,
If I'm afraid at all
It's only in my dreams.

I've got magic charm
That I keep up my sleeve
I can walk the ocean floor
And never have to breathe.

Life doesn't frighten me at all
Not at all
Not at all.

Life doesn't frighten me at all.

International Instruments: Conventions, Declarations, Resolutions and Frameworks

- Universal Declaration of Human Rights (1948)
  - http://www.un.org/rights/50/decla.htm
- Geneva Convention Relative to the Protection of Civilian Persons in the Time of War (1949)
  - http://www2.ohchr.org/english/law/civilianpersons.htm

See also:

- Protocol Additional to the Geneva Conventions, relating to the Protection of Victims of International Armed Conflicts (Protocol I) (1977)
  - http://www.icrc.org/ihl.nsf/7c4d08d9b287a42141256739003e 636b/f6c8b9fee14a77fdc125641e0052b079
- Protocol Additional to the Geneva Conventions, relating to the Protection of Victims of Non-International Armed Conflicts (Protocol II) (1977)

- http://www.icrc.org/ihl.nsf/7c4d08d9b287a42141256739003 e636b/d67c3971bcfficioci25641e0052b545
- Convention Relating to the Status of Refugees (1951)
  - http://www2.ohchr.org/english/law/refugees.htm
- UNESCO Convention Against Discrimination in Education (1960)
  - http://www2.ohchr.org/english/law/education.htm
- International Convention on the Elimination on All Forms of Racial Discrimination (1965)
  - http://www2.ohchr.org/english/law/cerd.htm
- International Covenant on Economic, Social and Cultural Rights (1966)
  - http://www2.ohchr.org/english/law/cescr.htm

See also:

- General Comment of the Committee on Economic, Social and Cultural Rights on 'The right to education (article 13)' (1999)
  - http://www.unhchr.ch/tbs/doc.nsf/0/ae1a0b126d068e868 025683c003c8b3b?Opendocument
- International Covenant on Civil and Political Rights (1966)
  - http://www2.ohchr.org/english/law/ccpr.htm
- Convention Concerning Minimum Age for Admission to Employment (1973)
  - http://www.ilo.org/ilolex/cgi-lex/convde.pl?C138
- Convention on the Elimination of All forms of Discrimination against Women (1979)
  - http://www2.ohchr.org/english/law/cedaw.htm
- Convention on the Rights of the Child (1989)
  - http://www2.ohchr.org/english/law/crc.htm
- Convention Concerning Indigenous and Tribal Peoples (1989)
  - http://www.ilo.org/ilolex/cgi-lex/convde.pl?C169
- Convention on the Protection of the Rights of All Migrant Workers and Members of Their Families (1990)
  - http://www2.ohchr.org/english/law/cmw.htm
- World Declaration on Education for All (1990)
  - http://www.unesco.org/education/efa/ed_for_all/background/ jomtien_declaration.shtml

- Rome Statue of the International Criminal Court (1998)
  - http://untreaty.un.org/cod/icc/statute/romefra.htm
- Guiding Principles on Internal Displacement (1998)
  - http://www.idpguidingprinciples.org
- Education for All Dakar Framework for Action (2000)
  - http://unesdoc.unesco.org/images/0012/001211/121147E.pdf
- United Nations Millennium Declaration (2000)
  - http://www.un.org/millennium/declaration/ares552e.pdf
- United Nations Security Council Resolution 1539 (2004)
  - http://www.un.org/Docs/sc/unsc_resolutions04.html
- United Nations Security Council Resolution 1612 (2005)
  - http://www.un.org/Docs/sc/unsc_resolutions05.htm
- United Nations Security Council Resolution 1674 (2006)
  - http://www.un.org/Docs/sc/unsc_resolutions06.htm
- Convention on the Rights of Persons with Disabilities (2008)
  - http://www.un.org/disabilities/convention/conventionfull.shtml
- Human Rights Council Resolution 8/4: The right to education (2008)
  - http://ap.ohchr.org/documents/E/HRC/resolutions/A_HRC_RES_8_4.pdf
- United Nations Security Council Resolution 1882 (2009)
  - http://www.un.org/Docs/sc/unsc_resolutions09.htm

Regional Instruments

- American Declaration of the Rights and Duties of Man (1948)
  - http://www.hrcr.org/docs/OAS_Declaration/Text/oasrights2.html
- European Convention for the Protection of Human Rights and Fundamental Freedoms (1950)
  - http://conventions.coe.int/treaty/en/treaties/Html/005.htm
- Protocol of San Salvador to the American Convention on Human Rights (1988)
  - http://www.oas.org/juridico/english/treaties/a-52.html

- African Charter on the Rights and Welfare of the Child (1990)
  - http://www.ilo.org/public/english/employment/skills/recomm/instr/afri_3.htm
- European Social Charter, Revised (1996)
  - http://conventions.coe.int/treaty/en/treaties/html/163.htm

Tools, Standards, and Key UN Reports

- The Code of Conduct for the International Red Cross and Red Crescent Movement and Non-Governmental Organisations (NGOs) in Disaster Relief (1994)
  - http://www.ifrc.org/publicat/conduct/
- Sphere, Humanitarian Charter and Minimum Standards in Disaster Response (2004)
  - http://www.sphereproject.org/
- The Inter-Agency Network for Education in Emergencies Minimum Standards for Education in Emergencies, Chronic Crises and Early Reconstruction (2004)
  - http://www.ineesite.org/standards
- FTI support to education in fragile states: A Progressive Framework (2007)
  - http://ineesite.org/uploads/documents/store/doc_1_Progressive_Framework.pdf
    With discussion document and guidelines (2008)
  - http://www.education-fast-track.org/library/FrameworkNOV04.pdf
- Report of the Special Rapporteur on the right to education: Right to education in emergency situations (2008)
  - http://www2.ohchr.org/english/issues/education/rapporteur/annual.htm
- Day of General Discussion of the Committee on the Rights of the Child: The right of the child to education in emergency situations (2008)
  - http://www2.ohchr.org/english/bodies/crc/docs/discussion/RecommendationsDGD2008.doc

Foreword / H.E. Miguel D'Escoto Brockmann

1. K. Cahill, *The Untapped Resource: Medicine and Diplomacy* (Maryknoll, N.Y.: Orbis, 1971).
2. K. Cahill, *Famine* (Maryknoll, N.Y.: Orbis, 1982).
3. K. Cahill, *Health and Development* (Maryknoll, N.Y.: Orbis, 1976).

2. Protecting Human Rights in Emergency Situations / Vernor Muñoz

1. Universal Declaration of Human Rights (adopted 10 December 1948; UNGA Res 217 A[III]).
2. For text of Dakar Framework of Action on Education for All, see http://www.unesco.org/education/efa/ed_for_all/dakfram_eng.shtml (accessed 27 April 2009).
3. Available online at http://www.un.org/millenniumgoals (accessed 27 April 2009).
4. For mandate and documents relating to the United Nations Special Rapporteur on Education, see http://www2.ohchr.org/english/issues/education/rapporteur/index.htm (accessed 27 April 2009).

5. According to the World Health Organization (WHO), the twentieth century was the most violent period in human history. E.G. Krug, *World Report on Violence and Health: Summary* (Geneva, 2002). Also available at http://www.who.int/entity/violence_injury_ prevention/violence/world_report/en/summary_en.pdf (accessed 27 April 2009).

6. M. Sinclair, *Planning Education in and After Emergencies* (Paris, 2002), available online at http://www.unesco.org/iiep/PDF/Fund73.pdf (accessed 27 April 2009).

7. Inter-Agency Network for Education in Emergencies (INEE), *Minimum Standards for Education in Emergencies, Chronic Crises and Early Reconstruction* (2004), available online at http://www.ineesite.org/index.php/post/inee_handbook (accessed 27 April 2009).

8. UNICEF, *The State of the World's Children* (New York, 2004), available online at http://www.unicef.org/publicaions/files/Eng_text.pdf (accessed 27 April 2009).

9. UNESCO, *EFA Global Monitoring Report: Gender and Education for All—The Leap to Equality* (2003–2004), available online at http://www.portal.unesco.org/education/en/ev .php-URL_ID = 23023&URL_DO = DO_TOPIC&URL_SECTION = 201.html (accessed 27 April 2009).

10. Women's Commission for Refugee Women and Children *Global Survey on Education in Emergencies* (New York, 2004), available online at http://www.reliefweb.int/rw/ lib.nsf/d6900sid/DPAL-5Y8H4D/$file/Ed_Emerg.pdf?openelement (accessed 27 April 2009).

11. See further, UNHCR, Report of Special Rapporteur on the Right to Education (2006), E/CN.4/2006/45.

12. Ibid., 9.

13. Ibid., 8.

14. B. O'Malley, *Education Under Attack: A Global Study on Targeted Political and Military Violence Against Education Staff, Students, Teachers, Union and Government Officials, and Institutions* (New York, 2007), available online at http://www.unesco.org/education/attack/ educationunderattack.pdf (accessed 27 April 2007).

15. R. Coomaraswamy, Statement at the Security Council Open Debate, http:// www.un.org/children/conflict/english/12-feb-2008-statement-at-the-security-council-open-deb.html (accessed 27 April 2009).

16. G. Machel, *Impact of Armed Conflict on Children* (1996) (A/51/306), available online at http://www.unifem.undp.org/machelrep.html (accessed 27 April 2009).

17. INEE, *Minimum Standards*.

18. International Covenant on Economic, Social and Cultural Rights (adopted 16 December, entered into force 3 January 1976 G.A. Res. 2200A (XXI), UN Doc. A/6316 (1966), 993 UNTS).

19. CRC, arts. 2 and 28.

20. UN Committee on the Rights of the Child, "General Comment 1: The Aims of Education" (2001), in "Note by the Secretariat, Compilation of General Comments and General Recommendations Adopted by Human Rights Treaty Bodies" HRI/GEN/1/Rev. 9 (Vol. II) CRC/GC/2001/1, available online at http://daccessdds.un.org/doc/UNDOC/ GEN/G08/422/43/PDF/G0842243.pdf.

21. Optional Protocol to the Convention on the Rights of the Child on the involvement of children in armed conflict (adopted 25 May 2000, entered into force 12 February 2002), GA Res. A/RES/54/262.

22. Protocols I and II of Geneva Conventions, the African Charter on the Rights and Welfare of the Child, the ILO Convention No. 182 concerning the prohibition and immediate action for the elimination of the worst forms of child labour, and the Rome Statute of the International Criminal Court also ban children recruitment.

23. See http://www2.ohchr.org/english/bodies/crc/comments.htm for details of the work of the CRC Committee (accessed 27 April 2009).

24. See http://www2.ohchr.org/English/bodies/crc/discussion2008.htm for details of Discussion Day and subsequent recommendations (accessed 27 April 2009).

25. Convention Relating to the Status of Refugees (adopted 28 July 1951, entered into force 22 April 1954), 189 UNTS 137.

26. The General Assembly progressively granted competence to OCHR on issues related to internally displaced populations, based on article 9 of the UNHCR statute.

27. Report of the Repesentative of the Secretary-General, Mr. Francis M. Deng, submitted pursuant to Commission resolution 1997/39: Guiding Principles on Internal Displacement (1998) DE/CN.4/1998/53/Add.2.

28. Geneva Convention Relative to the Protection of Civilian Persons in Time of War (adopted 12 August 1948, entered into force 21 October 1950, 75 UNTS 287) (GC IV).

29. See also Art. 50 of GC IV relating to obligations of occupying powers and Art. 78 of OP I to the protection of the right to education during evacuation.

30. Protocol II Additional to the Geneva Conventions (adopted 8 June 1977, entered into force 7 December 1978, UN Doc. A/32/144 Annex II, 1125 UNTS no. 17513).

31. Rome Statue of the International Criminal Court (adopted 17 July 1998, entered into force 1 July 2002, UN Doc. A/CONF. 183/9; 37 ILM 1002 (1998); 2187 UNTS 90).

32. Ibid., Art. 8, para. 2. iv.

33. For details of World Conference on Education for All, see http://portal.unesco.org/education/en/ev.php-URL_ID = 37612&URL_DO = DO_TOPIC&URL_SECTION = 201.html (accessed 27 April 2009).

34. For the text of the Dakar Framework of Action on Education for All, see http://www.unesco.org/education/efa/ed_for_all/dakfram_eng.shtml (accessed 27 April 2009).

35. For details of the World Education Forum, see http://www.unesco.org/education/efa/wef_2000 (accessed 27 April 2009).

36. Available online at http://www.un.org/millenniumgoals (accessed 27 April 2009).

37. See note 7.

38. See further http://www.ineesite.org (accessed 27 April 2009).

39. "Report by the Director-General on Global Action Plan to Achieve the Education for All Goals" (March 2006), Doc. 174 EX/ para. 8, available online at http://unesdoc.unesco.org/images/0014100144211442450.pdf (accessed 27 April 2009).

40. Canadian International Development Agency and the Inter-Agency Network for Education in Emergencies, *Policy Roundtable in Emergencies, Fragile States and Reconstruction: Addressing Challenges and Exploring Alternatives* (New York, 2006), 36, aailable online at http://www.ineesite.org/index.php/post/policy_roundtable_2006 (accessed 27 April 2007).

41. R. Winthrop and R. Mendenhall, *Education in Emergencies: A Critical Factor in Achieving the Millenium Development Goals* (Winnepeg, 2006).

42. M. McGillivray, *Aid Allocation and Fragile States* (London, 2005), 13–14.

42. Ibid.

43. Statement for endorsement by the UN Secretary-General, Symposium on Nutrition in the Context of Conflict and Crisis, (Standing Committee on Nutrition, Berlin, 2002), available online at http://www.unscn.org/archives/scnnews24/ch4.htm (accessed 27 April 2009).

44. The Inter-Agency Education Cluster on Education was created by the Working Group of the Interagency Standing Committee in 2006. For further information on the Inter-Agency Standing Committee, see http://www.humanitarianinfo.org/iasc/page loader.aspx (accessed 27 April 2009). For further information on the Inter-Agency Education Cluster, see http://www.humanitarianreform.org/humanitarianreform/Default .aspx?tabid = 115 (accessed 27 April 2009).

45. Mireille Affa'a Mindzie, *Children Associated with Armed Forces* (Cape Town, 2008).

46. Information gathered during a visit in my capacity as Special Rapporteur to the Ivory Coast, by invitation of Save the Children International Alliance.

47. UNCHR, "Report of the Special Rapporteur on Education on the right to education of persons with disabilities" (2007) A/HRC/4/29.

48. J Bernard, "With peace in mind: Assessment as a tool for cultivating the quality of education in emergencies and long term reconstruction" (UNESCO, Basic Education Division, 2008).

49. J. Dolan, *Last in Line, Last in School: How Donors Are Failing Children in Conflict-affected Fragile States* (London, 2007), available online at http://www.savethechildren.org .uk/en/docs/last_in_line_long.pdf (accessed 27 April 2009).

3. The Child Protection Viewpoint / Alec Wargo

1. For the purposes of this chapter, "education" is defined as the individual persons that make up an education system who can feasibly be engaged in protection activities, including but not limited to administrators, teachers, parent-teacher associations, and, with safeguards, children themselves.

2. Under Security Council resolutions 1612 (2005) and 1882 (2009), grave rights violations during armed conflict include recruitment or use of children by armed forces or groups, killing and maiming of children, sexual violence perpetrated against children, attacks on schools and hospitals, abduction of children and denial of humanitarian access for children. Of course, these are not exhaustive, but they are often indicative of a wider range of abuse suffered by children during wartime.

3. The UN-accepted definition of a child soldier is any child, girl or boy, recruited or used by an armed force or group in any capacity including, but not limited to, combatants, sexual slaves, porters, cooks, spies, and the like.

4. It is the viewpoint of the United Nations that children under legal age of recruitment are, by their very minority, unable to make a mature choice on what harm may befall them if they are recruited and used in armed conflict.

4. Donor Investment for Education in Emergencies / Brenda Haiplik

1. Current debate at the global level recognizes the problematic nature of mainly donor nations applying the term "fragility" to recipient countries that see themselves in a more positive way.

2. International Save the Children Alliance (ISCA), *Delivering Education for Children in Emergencies: A Key Building Block for the Future* (London: International Save the Children Alliance, 2008), p. 1. This is a key advocacy document explaining the rationale for investing in education in emergencies.

3. In internally displaced person (IDP) camps, education can provide valuable activities, including life skills and survival techniques, as well as help adolescents combat idleness and boredom, encouraging them to stay within the safe camp environment.

4. Brenda Haiplik, The Cluster System in Pakistan: Focus on the Education Sector. Oxford: Forced Migration Review. (November 2007). Within two weeks of the devastating Pakistan earthquake in October 2005, temporary learning spaces were set up. Parents felt comfortable with the knowledge that their children were in a safe and protective learning environment, allowing them to focus on key recovery activities such as searching for missing relatives and securing livelihoods.

5. M. Sinclair, *Planning Education in and After Emergencies* (Paris: IIEP-UNESCO, 2002), p. 27.

6. ISCA, *Delivering Education for Children in Emergencies*, p. 2.

7. L. Brannelly, S. Ndaruhutse, and C. Rigaud, *Donors' Engagement in Education in Fragile and Conflict-Affected States* (Berkshire: CBT Education Trust and UNESCO International Institute for Educational Planning [Paris], 2008), p. 2. A few donors are becoming strong advocates of funding education in emergencies. The EC included education in emergencies as one of the three strands within the action framework of its 2007 and 2008 operational strategy under the Children's sectoral theme. Education is at the center of Dutch development policy and is included as a key sector for humanitarian aid. The Dutch government leads the dialogue with Fast Track Initiative (FTI) partners in exploring appropriate financing options for education in fragile states through the FTI. The Netherlands' commitment of USD 201 million via UNICEF to support education in forty countries in crisis is noteworthy. Many donors now use a mix of aid

modalities, yet prefer pooled funding around sectorwide approaches and budget support when possible.

8. E. Waerum Rognerud, *Education in Emergencies and Reconstruction: Bridging the Funding Gap* (Paris: UNESCO International Institute for Educational Planning, 2005), p. 21.

9. UNICEF, *Education in Emergencies Training Resources* (Nairobi: Eastern and Southern Africa Regional Office [UNICEF], 2009).

10. The Inter-Agency Network on Education in Emergencies (INEE), Chronic Crises and Early Reconstruction Education and Fragility Working Group supports critical research on the economic value of investing in education in emergencies.

11. INEE, *Education in Emergencies: Including Everyone* (Geneva: INEE, 2009), p. 9.

12. ISCA, *Delivering Education for Children in Emergencies*, p. 4.

13. INEE, *Education in Emergencies*, p. 9.

14. The MS are as critically important in the pre-emergency preparedness phase as they are during all phases of an emergency. The MS provide a common framework for bringing together diverse needs, varied definitions, and numerous actors in the education sector.

15. Many alternative education programs include modules on peace education and social cohesion.

16. INEE, *Education in Emergencies*, p. 16.

17. In Sri Lanka (2009) UNICEF and Save the Children jointly lead the education cluster. Both organizations also sit on the Inter-Cluster Working Group representing the education sector.

18. The generous donation from the government of the Netherlands to UNICEF for education in emergencies has already made a significant impact at both school community and strategic policy levels in several countries.

19. Several excellent resources exist on psychosocial support during emergencies, including the IASC Guidelines on Mental Health and Psychosocial Support in Emergency Settings, providing useful guidance on how to strengthen access to safe and supportive education and support to and motivation of educators.

8. An Unexpected Lifeline / Gerald Martone

1. George Kent, *The Politics of Children's Survival* (New York: Praeger, 1991).

2. Rebecca Winthrop and Mary Mendenhall, *Education in Emergencies: A Critical Factor in Achieving the MDGs* (London: Commonwealth Ministers Reference Book, 2006).

3. World Health Organization, *World Report on Violence and Health: Summary* (Geneva: World Health Organization, 2002).

4. Graca Machel, *Promotion and Protection of the Rights of Children: Impact of Armed Conflict on Children* (New York: United Nations, 1996).

5. Susan Nicolai and Carl Triplehorn, *The Role of Education in Protecting Children* (New York: Humanitarian Practice Network, 2003).

6. Johann Pottier, "Agricultural Rehabilitation and Food Insecurity in Post-War Rwanda: Assessing Needs, Designing Solutions," *Institute of Development Studies Bulletin* 27, no. 3 (1996).

7. "The World's Unschooled," *New York Times*.

8. Janice Dolan, *Last in School, Last in Line: How Donors Are Failing Children in Conflict-affected Fragile States* (New York: International Save the Children Alliance, 2007).

9. Plan International, *Because I Am a Girl: The State of the World's Girls 2007* (London: Plan International, 2007).

10. "Educating Girls," *New York Times*, June 25, 2005.

11. Women's Refugee Commission, *Global Survey on Education in Emergencies: Women's Commission on Refugee Women and Children, 2006* (New York: Women's Refugee Commission, 2006).

12. Rebecca Winthrop and Jackie Kirk, "Teacher Development and Student Well-being," *Forced Migration Review*, January 2005.

13. David Werner and David Sanders, *Questioning the Solution: Politics of Primary Health Care and Child Survival* (New York: Health Wrights, 1997).

14. Eva Ahlen, "UNHCR's Education Challenges," *Forced Migration Review* 26 (July 2006).

15. World Health Organization, *The Invisible Wounds: The Mental Health Crisis in Afghanistan* (Geneva: WHO Special Report, Central Asia Crisis Unit, 2001).

16. Tony Vaux, *The Selfish Altruist: Relief Work in Famine and War* (London: Earthscan, 2001).

17. Michael Bryans, Bruce D. Jones, and Janice Gross Stein, *Mean Times: Humanitarian Action in Complex Political Emergencies—Stark Choices, Cruel Dilemmas* (Toronto: NGOs in Complex Emergencies Project, University of Toronto, 1999).

18. Mary B. Anderson, *Do No Harm: How Aid Can Support Peace—Or War* (Boulder, Colo.: Lynne Rienner, 1999).

19. Inter-Agency Network for Education in Emergencies, *Minimum Standards for Education in Emergencies, Chronic Crises, and Early Reconstruction* (Paris: UNESCO, 2004).

20. Mark Malloch Brown, "Refugees: The African Dimension," paper given at the Symposium on Assistance to Refugees: Alterative Viewpoints, Oxford, March 1984.

21. ALNAP, *Humanitarian Action: Improving Performance Through Improved Learning* (London: Overseas Development Institute, 2002).

22. Atuu Waonaje, speech at the Voices of Courage Award Ceremony, Women's Commission for Refugee Women and Children.

23. Women's Refugee Commission, *Global Survey*.

24. ALNAP, *Humanitarian Action*.

25. NRC/Save the Children Norway/UNHCR, Protection of Children and Adolescents in Complex Emergencies, conference and workshop, Oslo, November 9–11, 1998.

26. Rachel Brett and Irma Specht, *Young Soldiers: Why They Choose to Fight* (Boulder, Colo.: Lynne Reiner, 2004).

27. Pijko Rasanen et al., "Maternal Smoking During Pregnancy and Risk of Criminal Behavior Among Adult Male Offspring in the Northern Ireland 1966 Birth Cohort," *American Journal of Psychiatry* 156 (1999), Marc Morial, "Parents, Young Black Youth, and Higher Learning," MaximsNews.com, May 17, 2007.

28. Plan International, *Because I Am a Girl.*

29. Paul Collier and David Dollar, "Aid Allocation and Poverty Reduction," *European Economic Review* 46, no. 8 (2002): 1475–1500.

30. Werner and Sanders, *Questioning the Solution.*

31. Dolan, *Last in School, Last in Line.*

32. Women's Education and Fertility Behavior, Recent Evidence from the Demographic and Health Surveys, UN Document ST/ESA/SER.R/137 (1995), page 30; B. J. Wattenberg, "The Population Explosion Is Over," *New York Times*, November 23, 1997.

33. Ban Ki-Moon, UN Secretary-General, "Message on World Population Day," July 11, 2007.

34. Plan International, *Because I Am a Girl.*

35. K. Subbarao and Laura Raney, "Social Gains from Female Education," *Economic Development and Cultural Change* 44, no. 1 (1995): 105–128.

36. Gene Sperling, "Education Could Be America's Best Defense," *Financial Times*, July 22, 2004.

37. Peter W. Singer, *Children at War* (Berkeley: University of California Press, 2006).

38. Celia W. Dugger, "Three Democrats Suggest Plans for Education in Poor Nations," *New York Times*, May 2, 2007.

39. Statement for endorsement by the UN Secretary-General, Symposium on Nutrition in the Context of Conflict and Crisis, 12–13 March 2002, 24 SCN News, Standing Committee on Nutrition (July 2002).

40. Winthrop and Mendenhall, *Education in Emergencies.*

41. Dolan, *Last in School, Last in Line.*

9. The Power of the Curriculum / Falk Pingel

1. I follow here the terminology of Robert Fiala, "Educational Ideology and the School Curriculum," in Aaron Benavot and Cecilia Braslavsky, eds., *School Knowledge in Comparative and Historical Perspective* (Hong Kong: Springer, 2006), 15–34.

2. A classic on this topic in general is Frances Fitzgerald, *America Revised: History Schoolbooks in the Twentieth Century* (Boston: Little, Brown, 1979). See also Fitzgerald, 1979; see also Kyle Ward, *History in the Making: An Absorbing Look at how American History Has Changed in the Telling Over the Last 200 Years* (New York: New Press, 2006), and William E. Marsden, *The School Textbook: Geography, History and Social Studies* (London: Woburn Press, 2001).

3. For a comprehensive description of international textbook revision, see Falk Pingel, *UNESCO Guidebook on Textbook Research and Textbook Revision* (Hannover: Hahnsche Buchhandlung, 2009).

4. Pilvi Torsti, *Divergent Stories, Convergent Attitudes: A Study on the Presence of History, History Textbooks and the Thinking of Youth in Postwar Bosnia and Herzegovina* (Helsinki: Taifuuni, 2003).

5. Péter Radó, *Transition in Education: Policy Making and the Key Educational Policy Areas in the Central-European and Baltic Countries* (Budapest: Open Society Institute, Institute for Educational Policy, 2001).

6. For research on the textbook and curriculum reform with a focus on history education, see Augusta Dimou, ed., *"Transition" and the Politics of History Education in Southeastern Europe* (Göttingen: Vandenhoeck & Ruprecht, 2009).

7. This opposition has been addressed by Michael W. Doyle and Nicholas Sambalis, *Making War and Building Peace* (Princeton: Princeton University Press, 2006).

8. See Ruth Firer and Sami Adwan, *The Israeli-Palestinian Conflict in History and Civics Textbooks of Both Nations*, edited by Falk Pingel (Hannover: Verlag Hahnsche Buchhandlung, 2004). For alternative material that addresses the conflict directly, see the Dual Narrative Project developed by the Peace Research Institute for the Middle East (http://www.vispo.com/PRIME); the material has been written by teachers and university professors of both sides and will be published soon.

9. Takashi Yoshida, "Advancing or Obstructing Reconciliation? Changes in History Education and Disputes Over History Textbooks in Japan," in Elizabeth A. Cole, ed., *Teaching the Violent Past: History Education and Reconciliation* (Lanham, Md.: Rowman & Littlefield, 2007), 57–79; Ryota Nishino, "The Political Economy of the Textbook in Japan, with Particular Focus on Middle-School History Textbooks, ca. 1945–1995," *Internationale Schulbuchforschung/International Textbook Research* 30 (2008): 487–514.

10. For a comparative view, see Laura Hein and Mark Selden, eds., *Censoring History: Citizenship and Memory in Japan, Germany, and the United States* (Armonk, N.Y.: M. E. Sharpe, 2000).

11. World Bank, *Reshaping the Future: Education and Post Conflict Reconstruction* (Washington, D.C.: World Bank, 2005).

12. Ibid.

13. Pauline Rose and Martin Greeley, *Education in Fragile States: Capturing Lessons and Identifying Good Practice* (Sussex: DAC Fragile States Group, 2006).

14. The report has also been criticized in this regard as well by Harvey M. Weinstein, Sarah Warshauer Freedman, and Holly Hughson, "School Voices: Challenges Facing Education Systems After Identity-based Conflicts," *Education, Citizenship and Social Justice* 2 (2007): 41–71.

15. *INEE Minimum Standards Guidance Notes*, Chapter 20: Curriculum content and review processes, p. 15, http://www.iiep.unesco.org/capacity-development/technical-assistnce/emergencies-and-fragile-contexts/introduction/guidebook.html.

16. The same applies to students. Surveys have shown that even in countries with sophisticated education systems offering a wide range of teaching devices, pupils regard the textbook as the most reliable and "objective" transmitter of the truth; they trust in the book even more than in the teacher, who is more amenable to subjective emotions and judgments (Magne Angvik and Bodo von Borries, eds., *Youth and History: A Comparative European Survey on Historical Consciousness and Political Attitudes Among Adolescents* (Hamburg: Körber-Stiftung, 1997).

17. June Bam, "Negotiating History, Truth and Reconciliation and Globalization: An Analysis of the Suppression of Historical Consciousness in South African Schools as Case Study," *Mots Pluriels* 13 (2000), http://www.arts.uwa.edu.au/MotsPluriels/MP1300jb.html.

18. Marc Sommers and Peter Buckland (2004) *Parallel Worlds: Rebuilding the Education System in Kosovo* (Paris and Washington, D.C.: UNESCO, International Institute for Educational Planning, and World Bank, 2004).

19. Elizabeth Oglesby, "Historical Memory and the Limits of Peace Education: Examining Guatemala's Memory of Silence and the Politics of Curriculum Design," in Elizabeth A. Cole, ed., *Teaching the Violent Past: History Education and Reconciliation* (Lanham, Md.: Rowman & Littlefield, 2007), 175–202.

20. Rose and Greeley, *Education in Fragile States*.

21. Weinstein, Freedman, and Hughson, "School Voices."

22. Nemer Frayha, (2004) "Developing Curriculum as a Means to Bridging National Divisions in Lebanon," in Sobhi Tawil and Alexandra Harley, eds., *Education, Conflict and Social Cohesion* (Geneva: UNESCO International Bureau of Education, 2004), 159–205.

23. Héctor Lindo-Fuentes, "Balancing Memory and 'Culture of Peace': Writing a History Textbook in El Salvador after Civil War," *Internationale Schulbuchforschung/International Textbook Research* 21 (1999): 339–351.

24. Ana Obura, *Never Again: Educational Reconstruction in Rwanda* (Paris: UNESCO, International Institute for Educational Planning, 2003).

10. Attacks on Education / Brendan O'Malley

1. Clancy Chassay, "Acid Attacks and Rape: Growing Threat to Women Who Oppose Traditional Order," *The Guardian*, 22 November 2008; *Children and Armed Conflict*, Report of the Secretary-General, 26 March 2009, A/63/785-S/2009/158.

2. Brendan O'Malley, *Education under Attack: A Global Study on Targeted Political and Military Violence Against Education Staff, Students, Teachers, Union and Government Officials, and Institutions* (Paris: UNESCO, 2007).

3. Radhika Coomaraswamy, *Annual Report of the Special Representative of the Secretary-General for Children and Armed Conflict*, UNGA A/HRC/12/49, 30 July 2009.

4. *The Gaza Blockade: Children and Education Fact Sheet*, Association of International Development Agencies, 18 July 2009. Figures supplied by UNICEF, UNRWA, and MOEHE.

5. *Up in Flames: Humanitarian Law Violations and Civilian Victims in the Conflict Over South Ossetia*, Human Rights Watch report, January 2009; *The Impact of Russian Invasion on Georgian Educational System*; *Russian Invasion of Georgia, Facts and Figures*, 25 September 2008; and *Report of the Secretary-General on Children and Armed Conflict*, 26 March 2009, A/63/785-S2009/158, 11.

6. Margaret Besheer, "UN: There Are More Than 250,000 Child Soldiers Worldwide," Voice of America, 30 January 2008.

7. Brendan O'Malley, *Education under Attack 2010* (Paris: UNESCO 2010)

8. *Children and Armed Conflict*.

9. O'Malley, *Education under Attack 2010*.

10. Henry G. Jarecki and Daniela Zane Kaisth, *Scholar Rescue in the Modern World* (New York: Institute of International Education, 2009).

11. O'Malley, *Education under Attack 2010*.

12. Marit Glad, *Knowledge on Fire: Attacks on Education in Afghanistan—Risks and Mitigatory Measures*, produced by CARE with the assistance of Co-ordination of Afghan Relief/Organization for Sustainable Development and Research, on behalf of the World Bank and the Ministry of Education, September 2009.

13. Figures supplied by Executive District Officer, Elementary and Secondary Education, Swat.

14. "Maoist Diktat in Bihar: Send Kids to School," *Times of India*, 11 June 2009. In April Maoists had blown up a school mainly used by tribal children, who are among the most marginalized.

15. Interview with Brendan O'Malley, 27 August 2009.

16. "Thailand: Bomb Injures 14 in Thai Muslim South," AFP, 18 June 2007.

17. Fred van Leeuwen, General Secretary, Education International, Letter to President Jalal Talabani of Iraq, 16 November 2006.

18. Peter Beaumont, "Iraq's Universities and Schools Near Collapse as Teachers and Pupils Flee," *The Observer*, 5 October 2006.

19. Address by Nicholas Burnett, Assistant Director-General for Education, UNESCO, to the United Nations General Assembly Thematic Debate on Education in Emergencies, 18 March 2009, available at: http://unesdoc.unesco.org/images/0018/001812/181220e.pdf.

20. Brendan O'Malley, "Baghdad Battles for Better Education," *South China Morning Post*, 17 January 2009.

21. Glad, *Knowledge on Fire*.

22. Fazlul Haque, interviewed by Brendan O'Malley, April 2009.

23. Ibid.

24. Melinda Smith, "Schools as Zones of Peace: Nepal Case Study in Access to Education During Armed Conflict and Civil Unrest," UNESCO paper, August 2009, 6;

Melinda Smith, "Schools as Zones of Peace," presentation to INEE, Istanbul, April 2009.

25. Melinda Smith, "Schools as Zones of Peace: Nepal Case Study in Access to Education During Armed Conflict and Civil Unrest," UNESCO paper, August 2009, 7.

26. Melinda Smith, "Schools as Zones of Peace," presentation to INEE, Istanbul, April 2009.

27. Ibid.; Smith, "Schools as Zones of Peace," UNESCO paper, 8–9.

28. Address by Nicholas Burnett.

29. *Report of the Secretary-General on Children and Armed Conflict.*

30. Human Rights Watch, "Taking the Next Step: Strengthening the Security Council's Response to Sexual Violence and Attacks on Education in Armed Conflict," April 2009.

31. Ibid.

32. For the arguments in detail, see Gregory Raymond Bart, "The Ambiguous Protection of Schools Under the Law of War: Time for Parity with Hospitals and Religious Buildings," *Georgetown Journal of International Law* 40, no. 2 (Winter 2009).

33. Lothar Krappman, addressing the UN General Assembly debate on Education and Emergencies, 18 March 2009.

12. Establishing Safe Learning Environments / Simon Reich

1. *The Human Security Report 2005*, "The Changing Face of Global Violence" (Part 1), http://www.humansecurityreport.info/HSR2005_PDF/Part1.pdf; "The Assault on the Vulnerable" (Part 3), http://www.humansecurityreport.info/HSR2005_PDF/Part3.pdf.

2. Paul Lagasse, "Child-Friendly Spaces Provide Crucial Safe Zones for Refugee Kids," http://www.charity.org/news/Child-Friendly_Spaces_Provide_Crucial_Safe_Zones_for_Refugee_Kids.

3. Maureen McClure and Gonzalo Retamal, "The Goat Stays: Wise Investments in Future Neighbors," in Scott Gates and Simon Reich, eds., *Child Soldiers in the Age of Fractured States* (Pittsburgh: University of Pittsburgh Press, 2009).

4. This issue is discussed in greater detail in the introduction to ibid.

5. The conflicts examined in these studies consisted of the following: Angola (1999–2002), Burundi (1993–2006), DRC (1996–2008), Sudan (1983–2008), Sierra Leone (1994–2001), Liberia (1989–1997, 1999–2003), and Uganda (1994–2008).

6. Ford Institute for Human Security, "What Makes a Camp Safe? The Protection of Children from Abduction in Internally Displaced Persons and Refugee Camps," http://www.fordinstitute.pitt.edu/docs/23182ReportPR11.pdf. This study was funded by the Glyn Berry Program of the Department of Foreign Affairs and International Trade (DFAIT), Government of Canada; Ford Institute for Human Security, "Protecting Civilians: Key Determinants in the Effectiveness of a Peacekeeping Force," http://www

.fordinstitute.pitt.edu/newsarticles/Protecting_Civilians_Final_Report.pdf. This second report was funded by the US Institute of Peace. For a list of the twenty-seven factors examined, see "What Makes a Camp Safe?" p. 3.

7. Ford Institute for Human Security, "What Makes a Camp Safe?" p. 4.

8. See Howard Adelman, "Why Refugee Warriors Are Threats," *Journal of Conflict Studies* 18, no. 1 (Spring 1998); Aristide Zolberg, Astri Suhrke, and Sergio Aguayo, *Escape From Violence: Conflict and Refugee Movements in the Developing World* (New York: Oxford University Press, 1989).

9. Vera Achvarina and Simon F. Reich, "No Place to Hide: Refugees, Displaced Persons, and the Recruitment of Child Soldiers," *International Security* 31, no. 1 (Summer 2006): 127–164.

10. United Nations Resolution S/RES/1674 (2006); International Development Research Center, "The Responsibility to Protect: A Report of the International Commission on Intervention and State Sovereignty" (2006), http://www.iciss.ca/pdf/Commission-Report.pdf.

11. Although Sierra Leone is only one such case, much of the data for the other conflicts studied follows the same trend. Even a statistical analysis of 1,180 cases supported the same conclusion; most children are abducted from camps early in a conflict.

12. Ramesh Thakur and David M. Malone, "UN Peacekeeping: Lessons Learned?" *Global Governance* 2 (June 2000), http://www.articlearchives.com/international-relations/national-security-foreign-war/328227-1.html.

13. The Darfur Consortium, "Putting People First: The Protection Challenge Facing UNAMID in Darfur," (July 2008), http://www.darfurconsortium.org/darfur_consortium_actions/reports/2008/Putting_People_First_UNAMID_report.pdf.

14. United Nations Security Council, "Report of the Secretary-General on Children and Armed Conflict in the Democratic Republic of the Congo," S/2008/693 (November 2008), http://www.archiviodisarmo.it/siti/sito_archiviodisarmo/upload/documenti/99113_ONU_children_ad_war_RD_Congo.pdf.

15. United Nations, *Report of the Panel on United Nations Peace Operations, 2000* (August 21, 2000), A/55/305-S/2000/809 http://www.reliefweb.int/library/documents/PeaceKeeping.pdf

16. Paul Collier et al., *Breaking the Conflict Trap: Civil War and Development Policy* (New York: World Bank and Oxford University Press, 2003).

17. For an excellent discussion of this issue, see Christopher Blattman and Jeannie Annan, "The Consequences of Child Soldiering," Institute of Development Studies, Sussex University, http://www.hicn.org/papers/wp22.pdf.

13. Psychosocial Issues in Education / Arancha García del Soto

I thank Jenna Felz and Lynn Tesher ("Ariel's Legacy") for their very helpful comments on the draft of this manuscript.

1. Michael Wessells, "Do No Harm: Challenges in Organizing Psychosocial Support to Displaced People in Emergency Settings," *Refugee* 23, no. 1 (2008): 6–13.

2. Inter-Agency Network for Education in Emergencies (INEE), *INEE Minimum Standards Handbook*, 1st ed. (London: DS Print|Redesign, 2004).

3. Ibid.

4. Ibid.

5. Ibid.

6. Brian O'Malley, *Education Under Attack* (Geneva: UNESCO, 2007).

7. Inger Agger and Jadranka Mimica, *ECHO Psycho-social Assistance to the Victims of War in Bosnia-Hercegovina and Croatia: An Evaluation* (Zagreb: ECHO and European Community Task Force Psycho-Social Unit, 1996).

8. Inter-Agency Standing Committee (IASC), *IASC Guidelines on Mental Health and Psychosocial Support in Emergency Settings* (Geneva: IASC, 2007.

9. Psychosocial Working Group, *Psychosocial Intervention in Complex Emergencies: A Conceptual Framework* (London: PSWG, 2003).

10. Ibid.

11. A. M. Kos, *They Talk We Listen* (Ljubljana: Slovene Foundation, 1997), p. 101.

12. Ibid., p. 102.

13. Ibid., pp. 107–115.

14. N. Richman, *Helping Children in Difficult Circumstances* (New York: Save the Children Foundation, 1991); N. Richman, *Communicating with Children* (New York: Save the Children Foundation, 1993).

15. Kos, *They Talk We Listen*, p. 103.

16. Joanne V Loewy and Andrea Frisch Hara, eds., *Caring for the Caregiver: The Use of Music and Music Therapy in Grief and Trauma* (Silver Spring Md.: American Music Therapy Association, 2007).

17. Kos, *They Talk We Listen*, 117–129.

18. Ibid., p. 132.

19. PSWG, *Psychosocial Intervention in Complex Emergencies*.

20. Kos, *They Talk We Listen*, p. 99.

14. Education in the IDP Camps of Eastern Chad / Gonzalo Sánchez-Terán

1. Issa H. Khayar, Le refus de l'école: Contribution a l'étude des problèmes de l'éducation chez les Musulmans du Ouaddai (Librarie d'Amérique et d'Orient, 1975).

2. In December 2006 it was extremely difficult to obtain proper figures on schooling rates from the Department of Education in Goz Beida, but education officials and teachers who were interviewed accepted this rate.

3. While many of these schools had once existed but had subsequently been closed because the teacher had left or the communities were unable to sustain it,

documents of the Department of Education still showed them as functioning institutions.

4. According to Oxfam's report Paying for People (2007), 2 million more teachers must be recruited across the developing world to make education for all a reality. Aid donors are failing to plug the gap: only eight cents of each dollar that is donated are channeled into government plans that include the training and salaries of teachers and health workers.

5. The World Bank, which advocated the imposition of user fees in the 1980s and early 1990s, has since revised this position, at least in terms of its public messaging. It no longer supports user fees in education, and a growing number of governments that have received debt relief are using the proceeds to abolish fees, such as Zambia, which announced the end of user fees for its rural population in 2006.

6. According to the WHO, AIDS spreads twice as quickly among uneducated girls than among girls that have at least some schooling.

7. This could be the reason why educational projects are often underfinanced. In the 2008 Consolidated Appeal Process for Chad ($318 million), only 14 percent were funded.

8. It is important to notice that this happens in education much more than in any other sector such as food distribution or water and sanitation, where water points, for instance, are divided by areas in the site.

9. The benefits of schooling children in a refugee or IDP crisis are countless: keeping children busy allows other activities to take place, as, for example, it gives mothers time to take care of the house, shop, and prepare food.

10. As UNICEF points out, no country has reached sustained economic growth without achieving near-universal primary education.

15. Education as a Survival Strategy: Sixty Years of Schooling for Palestinian Refugees / Sam Rose

1. Photo-story available at: http://electronicintifada.net/v2/article10223.shtml.

2. Convention on Certain Conventional Weapons, Protocol III—Protocol on Prohibitions or Restrictions on the Use of Incendiary Weapons, Geneva, 10 October 1980.

3. UNRWA's General Fund Budget for education in 2009 was US$283,507,000. See UNRWA in figures, available at: http://www.un.org/unrwa/publications/pdf/uif-deco8.pdf.

4. Report of the Director of UNRWA (Paris: UNRWA, 1951).

5. Maya Rosenfeld, "From Emergency Relief Assistance to Human Development and Back: UNRWA and the Palestinian Refugees, 1950–2009," Refugee Studies Quarterly (forthcoming).

6. UNRWA: A Brief History, 1950–1982 (Vienna: UNRWA, n.d.).

7. Ibid.

8. B. Schiff, *Refugees Unto the Third Generation: UN Aid to Palestinians* (Syracuse, N.Y.: Syracuse University Press, 1995).

9. This approach finds voice in the "W" for works in UNRWA's name. The Economic Survey Mission, which led to UNRWA's establishment, estimated in 1949 that works projects could lead to the deletion of 100,000 names from UNRWA's ration rolls by mid-1951. The need to limit the Agency's relief budget was given added urgency by global flour shortages and the Korean War, which pushed up prices of nonfood items. See *Report of the Director of UNRWA*.

10. One of these projects involved the construction of a canal to divert water from the Suez Canal to the Sinai to support the economic development of the area and the resettlement of refugees. See http://untreaty.un.org/unts/1_60000/5/22/00009078.pdf. UNRWA also committed to advance US$83 million for the Aswan Dam project, although the initiative was eventually shelved.

11. UNESCO *Executive Board, Resolutions and Decisions Adopted By the Executive Board at Its 30th Session*. The statement was in response to a communiqué from UNRWA's Deputy Director that it would be unable to increase its budget for primary education, since this would not further the primary goal of making refugees economically independent.

12. This five-year plan included a proposal to establish a nine-year cycle of elementary and preparatory schooling, a system that continues to this day. See Friedhelm Ernst, "Problems of UNRWA School Education and Vocational Training," *Journal of Refugee Studies* 2, no. 1 (1989). It should also be noted that when the General Assembly extended UNRWA's mandate in 1959, it made explicit reference to the need for an expanded vocational training program.

13. UNRWA *Annual Report 1960*.

14. Ibid. In June 1963, Davis reported that these plans had been implemented successfully: the capacity of UNRWA's training centers had actually grown from around 600 to 4,000, while investment in general education programs had brought "educational opportunities available for refugee children almost up to the level of those which existed for the children of the local population in Jordan, Lebanon, the Syrian Arab Republic and the Gaza Strip." See *Yearbook of the United Nations, 1963*.

15. Rosenfeld, "From Emergency Relief Assistance to Human Development and Back."

16. See *Finding Means, UNRWA's Financial Crisis and Refugee Living Conditions* (New York: FAFO, 2003).

17. Schiff, *Refugees Unto the Third Generation*.

18. *Report of the Director of UNRWA*.

19. UNRWA *Annual Report, 1950*; Rosenfeld, "From Emergency Relief Assistance to Human Development and Back."

20. UNRWA *Annual Report, 1950–1951*; Nabil A. Badran, "The Means of Survival: Education and the Palestinian Community, 1948–1967," *Journal of Palestine Studies* 9, no.

4 (Summer 1980). By comparison, an estimated 97 percent of Jewish children in Mandate Palestine were in school by 1944. See Rosemary Sayigh, "Palestinian Education—Escape Route or Strait-Jacket?" *Journal of Palestine Studies* 14, no. 3 (Spring 1985).

21. Ernst, "Problems of UNRWA School Education and Vocational Training."

22. From UNRWA's perspective, investment in education programs also had tactical motivations. Perennially straitjacketed by resource constraints, the Agency saw education and training programs as a way to support refugee self-sufficiency and reduce dependency on monthly food rations.

23. Muhammad Hallaj, "The Mission of Palestinian Higher Education," *Journal of Palestine Studies* 9, no. 4 (Summer 1980).

24. For a detailed exposition of the concept of education as a "family project," see Maya Rosenfeld, *Confronting the Occupation* (Stanford: Stanford University Press, 2004).

25. Hallaj, "Mission of Palestinian Higher Education"; Sayigh, "Palestinian Education."

26. Hallaj, "Mission of Palestinian Higher Education." This arrangement continues to this day, in all fields except Lebanon.

27. UNRWA: A Brief History, 1950–1982.

28. UNRWA Medium Term Plan, 2004–2009.

29. Schiff, *Refugees Unto the Third Generation*.

30. *Memorandum submitted by UNRWA to the UK Treasury*, February 2006.

31. See UNRWA Annual Reports, 1967–1980.

32. To help counter the loss of educational services, UNRWA produced and distributed self-learning materials and audiovisual aids. Requests to Israel to permit the extension of the school year were often rejected, although the agency argued that such moves might reduce Intifada violence. See UNRWA Annual Reports, 1988–1993; B. Schiff, *Refugees Unto the Third Generation*.

33. This system had been a permanent feature of life for Palestinians in the occupied territories since the 1980s, with closures typically tightened or relaxed in response to political or security-related considerations.

34. See Bir Zeit University Right to Education Campaign, http://right2edu.bir zeit.edu.

35. Palestinian National Early Recovery and Reconstruction Plan for Gaza, March 2009.

36. Much of the criticism of Palestinian textbooks has been based on research published by the Center for Monitoring the Impact of the Peace (CMIP), now known as the Institute for Monitoring Peace and Cultural Tolerance in School Education (www .impact-se.org). In response to the allegations made by CMIP, a number of other studies have been commissioned, including by the U.S. Consulate in Jerusalem. Those involved in these studies include the Israel/Palestine Center for Research and Information (IPCRI), the Georg Eckert Institute and the Harry S. Truman Institute for the Advancement of Peace at the Hebrew University in Jerusalem. Nathan Brown, professor of political science at George Washington University, has also published extensively on

this subject. His analysis of the controversy is available at http://www.geocities.com/ nathanbrown1/Adam_Institute_Palestinian_textbooks.htm.

37. The reintroduction of school feeding programs, a staple in UNRWA schools in earlier decades, in Gaza in 2008 is a sign of the socioeconomic regression that plagues the territory.

38. See *West Bank and Gaza Education Sector Analysis: Impressive Achievements Under Harsh Conditions and the Way Forward to Consolidate a Quality Education System* (New York: World Bank, 2006).

39. I am grateful to Rema Hammami of the Institute of Women's Studies at Bir Zeit University for this analysis, and also for the insights of Penny Johnson.

40. Likewise, during the first Intifada, requests from UNRWA to Israel to permit the extension of the school year were often rejected, although the agency argued that such moves might reduce violence. See *UNRWA Annual Reports, 1988–1993*.

16. Education in Afghanistan: A Personal Reflection / Leslie Wilson

1. Deputy Director for Communication, Advocacy and Program Support: March 2003–June 2005; Country Director: July 2005–April 2009.

2. In this essay, I will comment on the period since mid-2002, and not the various past states of public education in Afghanistan: the royalist years through the early 1970s; the Soviet war era; the post–Soviet civil war years (1980s and early 1990s); and the time of Taliban hegemony. I hasten to acknowledge that the current state of educational affairs in the country is significantly shaped, in various and sundry ways, by all these past times.

3. When a child enrolls in a government school, she or he remains enrolled on the books for at least three years whether attending class or not. So, enrollment numbers are not a reliable measure of attendance and promotion.

4. Principally, the World Bank, the European Union and the US Agency for International Development (USAID), but other multilateral and bilateral donors as well.

5. Countless literacy as well as numeracy and life-skills programs for older girls and women as well as chronically unschooled older boys were, and continue to be, offered under the auspices of the Ministry of Education and other ministries as well. These are not considered in this essay; again, despite their import to the overall educational scene in Afghanistan.

6. There were four ministers between 2002 and 2009.

7. An issue for teachers and teaching in Afghanistan is the Ministry of Education's enduring inability to recognize, verify, and validate the experience and expertise, which is absolutely real, of teachers who have acquired and honed their knowledge and skills in systems outside Afghanistan, irrespective of the profound disarray and dilapidation of the country's own systems.

8. In fact, I question if many or most donors and advisers actually thought very often of the educational needs and learning capacities of Afghan children, specifically and situationally.

9. Without question, NGOs were not blanketing the entire country, or even the secure provinces, with education services. There was, and is still, a much disparity in coverage/access as with quality. Many provinces have entire districts simply do not have teachers and schools for their children, still.

10. The story of textbook development, vetting, printing, and distribution is complex and deserving of separate deep scrutiny, and I will mention it only cursorily in this chapter.

11. This document is subtitled, "For the Afghan National Development Strategy (With Focus on Priorities)."

12. Throughout my years in Afghanistan, I was continually amazed to hear that Afghans, almost instantaneously, came to view the presence of outsiders, especially Westerners from liberal democracies, as occupiers and not, by a long shot, liberators. There are deep-rooted historical reasons for this, of course. Nonetheless, I continue to be bemused by the notion that "the West" has occupier and/or conqueror interests in Afghanistan. Perhaps this speaks to my geopolitical naiveté, or perhaps to my humanitarian instincts, which I know I share with endless numbers of Afghan and international development workers across the decades and, I hope, which will endure well into the future, for the sake of Afghan children and families.

13. This means that the funds were not flowing to/through the government treasury; rather, they were being spent directly by the non-government entity: some foreign donor agencies and militaries. The implication is that a true accounting of exactly how much is being spent, and for what purpose(s), is impossible for the government to know and report.

14. For example, reduced mortality in children under the age of five and for women giving birth.

15. Ministry school classes, formal and nonformal education in community-based schools, home-based literacy classes, accelerated learning, and so forth.

16. I was always a bit confounded to work in an environment that I viewed, generally, as a reconstruction and development one (on the relief–reconstruction–development spectrum) while, in many important ways, Afghanistan was more (as I noted on page one) an ongoing, complex emergency. Yet, when a situation is not an acute one, where responses are formulaic and agreed by donors and humanitarian workers, programming is more challenging, to say the least.

17. Gradually since 2002, the humanitarian community, including many reputable NGOs and UN agencies, became increasingly adamant that the involvement of Provincial Reconstruction Teams, which were military units of variable purpose (depending on the hosting foreign nation) should absent themselves from the humanitarian and development sector, that is, from clinic and school construction, from pharmaceuticals and school supplies distributions, and so on. The community's voice was variably heard

and not entirely respected, but many recognized humanitarian actors remain unconvinced of the positive value, practical or propaganda, of military and quasi-military groups undertaking humanitarian, or even charity, activities.

18. Education Sector Strategy for the Afghanistan National Development Strategy (With Focus on Prioritization), March 2007, page 4.

19. I am of two minds on this point. For example, because I understand the power of the press, I assiduously sought a media outlet for the sorry years-long story of student textbooks gone missing from classrooms countrywide. Eventually, a very soft story appeared, which was a step in the right direction. Still, often, I wished for reports and human-interest stories of the good news that was every day evident to me.

20. When I spoke at the UN on March 18, 2009, one of my initial comments as I looked out at the General Assembly was that the very assembly was itself testimony to the undeniable value and imperative of education for all. To this point, I assert that it cannot possibly be lost on Afghanistan's economic, political, agricultural, civil society, public health, and social service leaders and their advisors that accessible quality basic education, at a minimum, is essential and, at the going rate, well past due for more attention than it has been accorded since 2002.

21. All textbook production for Afghan students to date has been funded by foreign donors, including Denmark and the United States, among others, and by individual NGOs with contributions from their own donors.

22. Formally, it is the Ministry of Labor, Social Affairs, Martyrs and Disabled; one of this ministry's responsibilities is oversight on government-sponsored kindergartens, which are most often associated with ministries themselves, providing daycare for government employees' children.

23. As noted earlier, these are quasi-military constructs associated with national foreign militaries that work in different ways throughout Afghanistan to variously keep/ensure peace, provide ambient security/safety and, in some nation's cases, involve themselves with humanitarian and/or charity work, which some characterize as development.

17. Child-Friendly: The Dual Function of Protection and Education in Myanmar / Ni Ni Htwe and Makiba Yamano

1. Tripartite Core Group, 2008, Post Nargis Joint Assessment Report.

2. Ibid.

3. The Education Cluster took the lead on Disaster Preparedness and Response during the recovery phase, which is a crucial part of education after a natural disaster. For details, please visit the Humanitarian Information Center for Myanmar website: http://myanmar.humanitarianinfo.org/education/default.aspx.

4. In Myanmar, the Ministry of Social Welfare oversees the protection of vulnerable groups, nonformal education, and early childhood care and development programs. The Ministry of Education is responsible for formal school curriculum and infrastructure.

5. The National Plan of Action was officially approved in July 2008.

6. The government of Myanmar ratified the UN Convention of the Rights of the Child in 1991 and enacted the National Child Law in 1993.

7. As of October 31, 2008.

8. As of May 2009.

9. People fled to monasteries, schools, and hospitals after the cyclone until they started rebuilding their own temporary shelters. These public places were crowded by displaced families, and had higher risk of abuse and exploitation for women and children.

10. Among children participated in the WV's CFS, there was no severe trauma case observed.

11. End of Program Evaluation Report, TANGO International, April 2009.

12. There was a relatively low level of agreement with the statement, "It is wrong to punish a child by hitting him or her" (65.5 percent of all households), due to the widespread practice of corporal punishment. Despite the strong awareness-raising and promotion of positive discipline, many parents still believe corporal punishment is acceptable for disciplining children.

18. After the Storm: Minority School Development in New Orleans / Juan Rangel

District Composite Report, Orleans Parish, 2004–2005, http://www.louisianaschools.net/lde/pair/DCR0405/DCR036.pdf

2. Mike Waller, "BESE Begins to Pave Way for Takeovers: Panel OKs Plan for Applications to Run 14 Failing N.O. Schools," *Times-Picayune*, January 14, 2004.

3. Recovery School District, "Recovery School District at a Glance," http://www.rsdla.net/InfoGlance.aspx (accessed 8–27–09).

4. Ibid.

19. The Sudan: Education, Culture, and Negotiations / Francis M. Deng

1. Roberta Cohen and Francis M. Deng, *Masses in Flight: The Global Crisis of Internal Displacement* (Washington D.C.: Brookings Institution Press, 1998), p. 111.

2. E. Mooney and C. French, "Barriers and Bridges: Access to Education for Internally Displaced Children," http://www.Brookings.edu/papers/2005/0111humanrights_mooney-aspx.

3. Cohen and Deng, *Masses in Flight*, p. 112.

4. Ibid.

5. Ibid.

6. The material in this section relies heavily on a book I wrote with Larry Minear, *The Challenges of Famine Relief: Emergency Operations in the Sudan* (Washington, D.C.: Brookings Institution Press, 1992).

7. Alex de Waal, *Famine That Kills: Darfur, Sudan, 1984–1985* (Oxford: Clarendon Press, 1989), p. 196.

8. Deng and Minear, *The Challenges of Famine Relief*, p. 34.

9. Quoted in Larry Minear, *Humanitarianism Under Seige: A Critical Review of Operation Lifeline Sudan* (Trenton, N.J.: Red Sea Press, 1991), p. 39.

10. Quoted in Deng and Minear, *The Challenges of Famine Relief*, p. 36.

11. Minear, *Humanitarianism Under Seige*, p. 65.

12. Ibid.

13. Message from the U.N. Secretary-General, H.E. Kofi Annan, in Kevin M. Cahill, *Traditions, Values and Humanitarian Action* (New York: Fordham University Press, 2003), xiii.

14. Ibid.

15. For the source of the material in this section, see Francis M. Deng, "Learning in Context: An African Perspective," in Alan Thomas and Edward W. Ploman, eds., *Learning and Development: A Global Perspective* (Toronto: Ontario Institute for Studies in Education, 1986), 91. See also Francis M. Deng, *Africans of Two Worlds: The Dinka in Afro-Arab Sudan* (New Haven: Yale University Press, 1978), and *Dinka Cosmology* (London: Ithaca Press, 1980).

16. Deng, *Africans of Two Worlds*, p. 71; and Francis Mading Deng, *The Man Called Deng Majok: A Biography of Power, Polygyny and Change* (New Haven: Yale University Press, 1986), p. 30, reprinted by the Red Sea Press, 2009.

17. Godfrey Lienhardt, *Divinity and Experience: The Religion of the Dinka* (Oxford: Clarendon Press, 1961), p. 26. See also Deng, *The Man Called Deng Majok*, p. 26.

18. Brenda Seligman and Charles G. Seligman, *The Pagan Tribes of the Nilotic Sudan* (London: Routledge and Sons Ltd., 1932), p. 178.

19. Godfrey Lienhardt, *Divinity and Experience*, p. 46.

20. Ibid, pp. 46–47.

21. Francis Madeng Deng, *Africans of Two Worlds*, p. 66; Francis M. Deng, "A Cultural Approach to Human Rights Among the Dinka," in Abdullahi's Ahmed An-Na'im and Francis M. Deng, eds., *Cross-Cultural Perspectives on Human Rights* (Washington D.C.: Brookings Institution Press).

22. G. Lienhardt, "The Western Dinka," in John Middleton and David Tait, eds., *Tribes Without Rulers* (New York: Oxford University Press, 1958), pp. 106–107.

23. Father A. Nebel, *Dinka Dictionary*, 1954, p. 315.

24. Deng, *Dinka Cosmology*, p. 58, quoted in Francis M. Deng, *War of Visions: Conflict of Identities in the Sudan* (Washington, D.C.: Brookings Institution Press, 1995), p. 196.

25. Deng, *Dinka Cosmology*, p. 42; Deng, *War of Visions*, p. 197.

26. Major Court Treat, *Out of the Beaten Track* (New York: E. P. Dutton, 1931), pp. 115–116.

27. Major Thitherington, "The Raik Dinka," *Sudan Notes and Records* 10 (1927): 159.

28. Quoted in Deng, *War of Visions*, p. 282.

29. G. Lienhardt, "Man in Society," in *The Listener* (London: BBC, 1963), p. 828. See also Lienhardt, *Divinity and Experience*, p. 248.

30. Quoted in the introduction to Francis M. Deng, *Self-determination and National Unity: A Challenge for Africa* (Trenton, N.J.: Red Sea Press, 2009). The material in this section builds on the introduction to Francis M. Deng, *Identity, Diversity and Constitutionalism in Africa* (Washington, D.C.: U.S. Institute of Peace, 2008).

31. Francis M. Deng, *Talking It Out: Stories in Negotiating Human Relations* (London: Kegan Paul International, 2007).

32. The title of an article by the author in Abdel Ghaffar M. Ahmed and Gunar Sorbo, *Management of the Crisis in the Sudan* (Bergen, Norway: Bergen University Centre for Development Studies, 1989).

33. I. William Zartman, *Ripe for Resolution* (New York: Oxford University Press, 1985).

ALLISON ANDERSON is the Director at INEE. Previously she served as INEE Coordinator for Minimum Standards.

MAYA ANGELOU is an internationally renowned poet, author, filmmaker, and civil rights activist. She has been awarded the Presidential Medal of Arts.

H.E. MIGUEL D'ESCOTO BROCKMANN was the President of the Sixty-third Session of the UN General Assembly. A Maryknoll priest, he has been a missionary in Latin America, the editor-in-chief of Orbis Books, and the Foreign Minister of Nicaragua.

KEVIN M. CAHILL, M.D., is Chief Advisor for Humanitarian and Public Health issues to the President of the United Nations General Assembly and is University Professor and Director of the Institute of International Humanitarian Affairs (IIHA) at Fordham University.

ROBERT COLES, M.D., is the author of numerous books, including *Children of Crisis*, which won the Pulitzer Prize. He has been a distinguished Professor of Child Psychiatry at Harvard and Duke Universities. He has been awarded both the Presidential Medal of Freedom and the National Humanities Medal.

FRANCIS M. DENG is the UN Undersecretary General for the Prevention of Genocide. He previously served as Undersecretary General for Internally Displaced Persons and has been a Professor at the City University of New York, Harvard and the Brookings Institute. He has also served as Ambassador and Secretary of State for Foreign Affairs of the Sudan.

ZLATA FILIPOVIĆ is the author of *Zlata's Diary*, a personal account depicting the horrors of the war in Bosnia, and the coeditor of *Stolen Voices: Young People's War Diaries, from WWI to Iraq*.

ARANCHA GARCÍA DEL SOTO was the Helen Hamlyn Senior Fellow at Fordham University's IIHA. She has worked, published, and taught in Europe, Sri Lanka, Africa, Latin America, and the United States on psychosocial interventions with survivors of violence.

BRENDA HAIPLIK is a Senior Specialist for Education at Save the Children. She had previously worked for UNICEF in Pakistan, Somalia, Kenya, and Bangladesh.

JENNIFER HOFMANN is the INEE Coordinator for Minimum Standards.

NI NI HTWE is the Project Coordinator of the Child Development and Protection Project for World Vision Myanmar.

GERALD MARTONE is the Director of Humanitarian Affairs at the International Rescue Committee.

H.H. SHEIKHA MOZAH BINT NASSER AL MISSNED is the UNESCO Special Envoy on Basic and Higher Education. She also serves as the Chairperson of the Qatar Foundation for Education, Science and Community Development.

VERNOR MUÑOZ is the UN Special Rapporteur on the Right to Education. He is the Director of the Human Rights Education Department at the Costa Rican Ombudsman Office. He is author of many books, articles, and studies on literature and human rights.

H.E. PIERRE NKURUNZIZA is the President of Burundi.

BRENDAN O'MALLEY is author of *Education under Attack*, a UNESCO study on targeted political and military violence against education staff, students, teachers, union and government officials, and institutions. He has been an international editor of the *Times Educational Supplement*.

FALK PINGEL is the past Director of the George Eckert Institute in Germany. He has advised UNESCO, OSCE, and numerous international NGOs and governments on curriculum and textbook content.

JUAN RANGEL is Chief Officer of the United Neighborhood Organization (UNO).

SIMON REICH is the Chairman of the Global Affairs Division at Rutgers University. He had been Director of the Ford Institute of Human Security and Professor of International Affairs at the University of Pittsburgh.

SAM ROSE IS Coordinator for Emergency Programs at UNRWA in Jerusalem.

GONZALO SÁNCHEZ-TERÁN is the Africa Editor for *Exterior Politica*. He also worked with the Jesuit Refugee Service organizing and managing emergency projects in Guinea, Liberia, Ivory Coast, and eastern Chad.

ALEC WARGO is Special Assistant to the SRSG/CAAC. He previously worked with OSCE and UNHCR in child protection in East Timor, Guinea and Bosnia.

LESLIE WILSON is the Country Director for Save the Children in Afghanistan. Formerly, she was Country Representative for Bangladesh.

MAKIBA YAMANO is the Protection Coordinator for World Vision, Myanmar.

The Center for International Humanitarian Cooperation and the Institute for International Humanitarian Affairs

The Center for International Humanitarian Cooperation (CIHC) is a public charity founded in 1992 by a small group of international diplomats and physicians who believed that health and other humanitarian endeavors sometimes provide the only common ground for initiating dialogue, understanding, and cooperation among people and nations shattered by war, civil conflicts, and ethnic violence. The Center has sponsored symposia, produces occasional papers, and has published the International Humanitarian Books Series of Fordham University Press. Titles in that series include *Basics of Humanitarian Missions; Emergency Relief Operations; Traditions, Values and Humanitarian Actions; Human Security For All: A Tribute to Sergio Vieira de Mello; Technology for Humanitarian Action; Tropical Medicine: A Clinical Text; To Bear Witness: A Journey of Healing and Solidarity;* and *The Pulse of Humanitarian Assistance.* Five of these volumes have been published in French by Robert Laffont.

The CIHC and its Directors were deeply involved in trying to alleviate the wounds of war in Somalia and the former Yugoslavia. A CIHC amputee center in northern Somalia was developed as a model for a simple, rapid, inexpensive program that could be replicated in other war zones. In the former Yugoslavia the CICH was active in prisoner and hostage release, in legal assistance for human-rights and political-rights violations, and facilitated discussions between combatants.

The Center directs the International Diploma in Humanitarian Assistance (IDHA) in partnership with Fordham University in New York, the United Nations System Staff College, and The Royal College of Surgeons in Ireland. It has offered IDHA programs in Egypt, Kenya, Ireland, Switzerland, and the United States. The CIHC also offers specialized training courses in humanitarian negotiations, international human rights, mental health in conflicts, and other relevant topics; these courses have been held in Malaysia, Sudan, Nicaragua, Spain, Ireland, Turkey, Hungary, England, and the United States. There are more than 1,400 graduates from 127 nations. They represent all agencies of the United Nations and most major nongovernmental organizations (NGOs) around the world. The CIHC has also provided staff support in recent years in crisis management in Iraq, East Timor, Aceh, Kosovo, Palestine, Albania, Pakistan, and other disaster or conflict zones.

The CIHC is closely linked with Fordham University's Institute of International Humanitarian Affairs (IIHA). The Directors of the CIHC serve as the Advisory Board of the IIHA. The President of the CIHC is the University Professor and Director of the Institute. CIHC officer Larry Hollingworth is Humanitarian Programs Director for the Institute. Peter Hansen, a Director of the CIHC, is Diplomat in Residence at Fordham.

Founded in December 2001, the Institute provides a vehicle through which the field of humanitarian assistance may interact with local, national, and international academic communities. Designed to forge partnerships with relief organizations it enables humanitarian field workers to develop collaborative relationships with Fordham University, and with the broad international community in New York City.

The IIHA offers an academic base for the study and development of international health, human rights, and other humanitarian issues, especially those that occur in periods of conflict. At a time when terrorism and war are at the forefront of world affairs, the IIHA trains humanitarian workers in the critical skills needed to aid in crises situations. The IIHA identifies fundamental needs and uses its talents, contacts, and resources to define practical solutions.

As part of its multidisciplinary training in humanitarian assistance, the IIHA offers both graduate and undergraduate courses as well as a

Masters Degree in Humanitarian Action. Through these, the IIHA creates a unique bridge between the Fordham community and frontline humanitarian action.

For more information, please see the following websites:

- www.cihc.org
- http://www.fordham.edu/academics/programs_at_fordham_/
  international_humani